HPLC Method
for
Determination of APIs
in
pharmaceutical formulation

By

Parimal M Chatrabhuji, Ph D

Chintan V Pandya, Ph D

Mukesh C Patel Ph D NET MBA

Title	HPLC Method for Determination of APIs in pharmaceutical
Author	Dr Parimal M. Chatrabhuji, Ph D Dr Chintan V Pandya, Ph D Dr Mukesh C Patel Ph D NET MBA
Published by	**Lulu Press, Inc.** 3101 Hillsborough Street Raleigh, NC 27607 North Carolina, United States
ISBN	978-1-329-07286-2
Copyright	©2002-2014 Lulu Holdings, Inc. and/or Lulu Press, Inc. All rights reserved. No part of this book may be reproduced or transmitted in any form or by any means, electronic or mechanical, including photocopying, recording or by any information storage and retrieval system, without written permission from the author, except for the inclusion of a brief quotation in review.

First Edition, 2015

Contents

ABOUT THE AUTHORS

Dr. Parimal M. Chatrabhuji did his Graduation from Bahauddin Science College, Junagadh, Saurastra University in 1993 and Post-Graduation from Department of Chemistry, Maharaja Krishnakumarsinhji Bhavnagar University, Bhavnagar in 1995. He did his Ph. D. under the guidance of Retd. Prof.(Dr) N. K. Undavia from Maharaja Krishnakumarsinhji Bhavnagar University, Bhavnagar on the topic ―A Study of Heterocyclic Compounds as Anti-Microbial Agents‖ in 2000. His 114 newly synthesized organic compounds have been screened for anticancer and/or anti HIV activities by National Cancer Institute and National Institute of Health, a US Government organization, Maryland, USA.

Presently, Dr.Chatrabhuji is actively engaged in teaching at P.S. Science and H.D.Patel Arts College – Kadi, affiliated with Hemchandracharya North Gujarat University as Assistant Professor. He has 10 years of teaching experience Under – Graduate classes at and 04 years of teaching experience of Post – Graduate classes. He is a recognized Ph.D. supervisor in three different Universities and presently 02 Ph.D. research scholars under his supervision.

Dr.Chatrabhuji has published 33 research papers in reputed journals and 04 books with International and national publishers. He is also a Member Board of Studies U.G. and P.G. (Chemistry Subject) Kadi Sarva Vishwavidyalaya – Gandhinagar and Member Executive Editorial Board in ―Journals Club for Applied Chemistry‖ (JCAC). He has also rendered expert services in different National workshops and seminars. Dr.Chatrabhuji has completed 01 research projected granted by University Grants Commission and one ongoing research project sponsored by Kadi Sarva Vishvavidyalaya, Gandhinagar.

Dr Mukesh Patel is distinguished faculty (Physical Chemistry) at PS Science and H D P Arts College, Kadi, affiliated to HNG University, India for more than two decades. He did PhD in 2002 and PG in 1994 and qualified UGC-CSIR NET for lectureship twice. He has completed three research projects sponsored by UGC. He has published more than 50 research papers and supervised 6 PhD scholars.

Dr Chintan Vijaykumar Pandya is an Asst. Professor in Analytical Chemistry presently teaching in post graduate level at HVPGR, Kadi – KSV (India). Dr Pandya has an excellent track record in academic institution of high repute, in research with more than ten international research articles & published five international books at very young age of 30 years and presents his work in number of national and international conferences.

PREFACE

The idea for writing this basic HPLC book was probably born during the project sanctioned by University Grants Commission – Pune. This book was written as an updated reference guide for busy laboratory analysts and researchers. Topics covered include HPLC operation, method development and validation aspects. This book can serve as a supplementary text for students pursuing a career in analytical chemistry. It describes basic theories and terminologies for the novice and reviews relevant concepts, best practices, and modern trends for the experienced practitioner. While broad in scope, this book focuses on reversed-phase HPLC (the most common separation mode) and pharmaceutical applications (the largest user segment). Information is presented in a straight forward manner and illustrated with an abundance of diagrams, chromatograms, tables, and case studies and supported with selected key references or web resources. The book covers a broad-scope overview of basic principles, instrumentation, and applications, concise review of concepts and trends relevant to modern practice. A reader with a science degree and a basic understanding of chemistry is assumed.

- Authors

1. Importance of drug analysis

'Health is wealth'. It is vital fact that a healthy body is desire of every human being. Good health is first condition to enjoy the life and all other things which mankind is having. Nowadays peoples are more concentrating towards health. Even governmental bodies of different countries and World health organization (WHO) are also focusing for health of human being. Health care is prevention, treatment and management of illness and preservation of mental and physical well-being. Health care embraces all the goods and services designed to promote health including preventive, curative and palliative in interventions. The Health care industry is considered an industry or profession which includes people's exercise of skill or judgment or providing of a service related to the prevention or improvement of the health of the individuals or the treatment or care of individuals who are injured, sick, disabled or infirm. The delivery of modern health care depends on an interdisciplinary team.

The medical model of health focuses on the eradication of illness through diagnosis and effective treatment. A traditional view is that improvement in health results from advancements in medical science. Advancements in medical science bring varieties of medicines. Medicines are key part of the health care system. The numerous medicines are introducing into the world-market and also, that is increasing every year. These medicines are being either new entities or partial structural modification of the existing one. So, to evaluate quality and efficacy of these medicines is also important factor. Right from the beginning of discovery of any medicine quality and efficacy of the same are checked by quantification means. Quality and efficacy are checked by either observing effect of drug on various animal models or analytical means. The option of animal models is not practically suitable for every batch of medicine as it's require long time, high cost and more man-power. Later option of analytical way is more suitable, highly precise, safe and selective.

The analytical way deals with quality standards which are assigned for products to have desirable efficacy of the medicines. Sample representing any batch are analyzed for these standards and it is assumed that drug/medicine which is having such standards are having desire effect on use. Quality control is a concept, which strives to produce a perfect product by series of measures designed to prevent and eliminate errors at different stage of production. The decision to release or reject a product is based on one or more type of control action.

Due to rapid growth of pharmaceutical industry during last several years, number of pharmaceutical formulations are enter as a part of health care system and thus, there has been rapid progress in the field of pharmaceutical analysis. Developing analytical method for newly introduced pharmaceutical formulation is a matter of most importance because drug or drug combination may not be official in any pharmacopoeias and thus, no analytical method for quantification is available. To check the quality standards of the medicine various analytical methods are used. Modern analytical techniques are playing key role in assessing chemical quality standards of medicine. Thus analytical techniques are required for fixing standards of medicines and its regular checking. Out of all analytical techniques, the technique which is widely used to check the quality of drug is known as 'CHROMATOGRAPHY'.

2. History of chromatography and HPLC

In 1903 a Russian botanist Mikhail Tswett produced a colorful separation of plant pigments through calcium carbonate column. Chromatography word came from Greek language chroma = color and graphein = to write i.e. color writing or chromatography [1, 2].

Prior to the 1970's, few reliable chromatographic methods were commercially available to the laboratory scientist. During 1970's, most chemical separations were carried out using a variety of techniques including open-column chromatography, paper chromatography, and thin-layer chromatography. However, these chromatographic techniques were inadequate for quantification of compounds and resolution between similar compounds. During this time, pressure liquid chromatography began to be used to decrease flow through time, thus reducing purification times of compounds being isolated by column chromatography. However, flow rates were inconsistent, and the question of whether it was better to have constant flow rate or constant pressure was debated [3].

High pressure liquid chromatography was developed in the mid-1970's and quickly improved with the development of column packing materials and the additional convenience of online detectors. In the late 1970's, new methods including reverse phase liquid chromatography allowed for improved separation between very similar compounds. By the 1980's HPLC was commonly used for the separation of chemical compounds. New techniques improved separation, identification, purification and quantification far above the previous techniques. Computers and automation added to the convenience of HPLC. Improvements in type of columns and thus reproducibility were made as such terms as micro-column, affinity columns, and Fast HPLC began to immerge. By the 2000 very fast development was undertaken in the area of column material with small particle size technology and other specialized columns. The dimensions of the General Introduction typical HPLC column are 100-300 mm in length with an internal diameter between 3-5 mm. The usual diameter of micro-columns, or capillary columns, ranges from 3 µm to 200 µm[4]. In this decade sub 2 micron particle size technology (column material packed with silica particles of < 2µm size) with modified or improved HPLC instrumentation becomes a popular with different instrument brand name like UPLC (Ultra Performance Liquid Chromatography) of Waters and RRLC (Rapid Resolution Liquid Chromatography) of Agilent.

3. Modern High Performance Liquid Chromatography (HPLC)

The highly sophisticated reliable and fast liquid chromatographic (LC) separation techniques are become a requirement in many industries like pharmaceuticals, agrochemicals, dyes, petrochemicals, natural products and others. Early LC used gravity fed open tubular columns with particles 100s of microns in size; the human eye was used for a detector and separations often took hours (days?) to develop.

Isocratic and Gradient LC System Operation

Two basic elution modes are used in HPLC. The first is called isocratic elution. In this mode, the mobile phase, either a pure solvent or a mixture, *remains the same throughout the run.* A typical system is outlined in Figure 1.

A schematic instrumentation of HPLC is given through figure 1-3 as under:

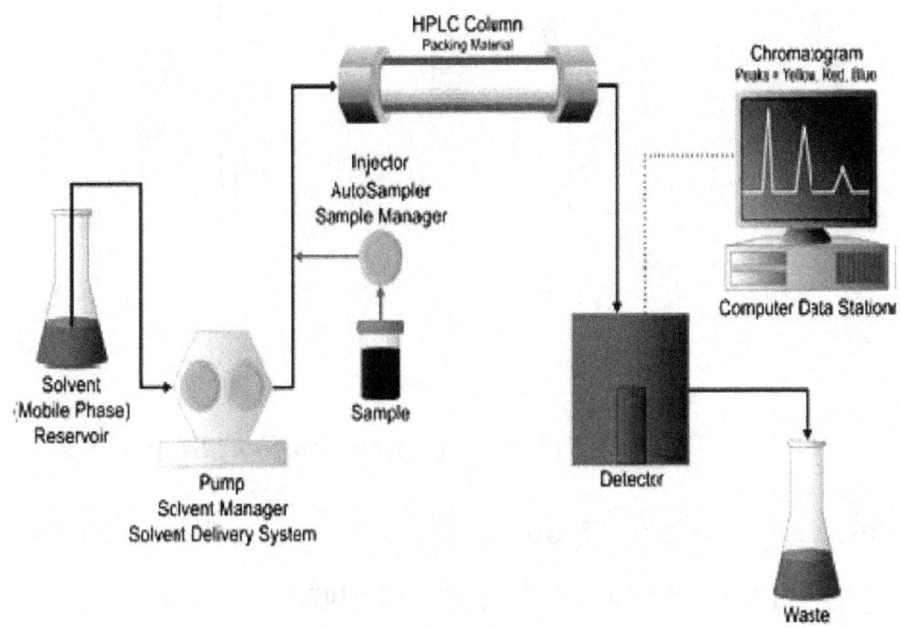

Figure - 1 Isocratic LC system

The second type is called gradient elution, wherein, as its name implies, *the mobile phase composition changes during the separation.* This mode is useful for samples that contain compounds that span a wide range of chromatographic polarity. As the separation proceeds, the elution strength of the mobile phase is increased to elute the more strongly retained sample components.

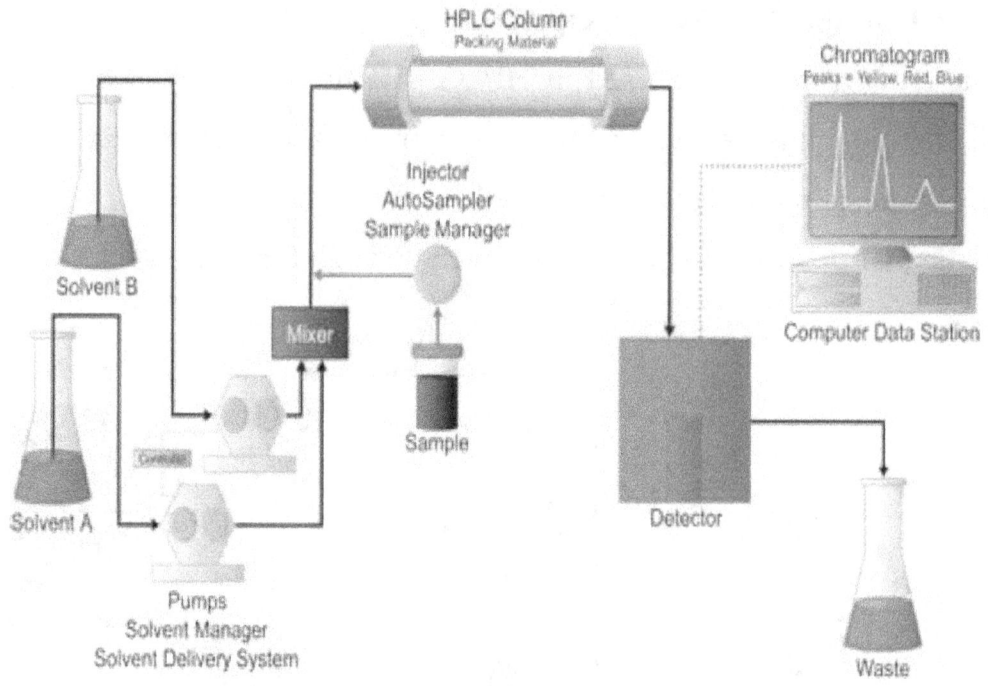

Figure – 2 High-Pressure-Gradient System

In the simplest case, shown in Figure 2, there are two bottles of solvents and two pumps. The speed of each pump is managed by the gradient controller to deliver more or less of each solvent over the course of the separation. The two streams are combined in the mixer to create the actual mobile phase composition that is delivered to the column over time. At the beginning, the mobile phase contains a higher proportion of the weaker solvent [Solvent A]. Over time, the proportion of the stronger solvent [Solvent B] is increased, according to a predetermined timetable. Note that in Figure 2, the mixer is downstream of the pumps; thus the gradient is created under *high pressure*. Other HPLC systems are designed to mix multiple streams of solvents under *low pressure*, ahead of a single pump. A gradient proportioning valve selects from the four solvent bottles, changing the strength of the mobile phase over time [Figure 3].

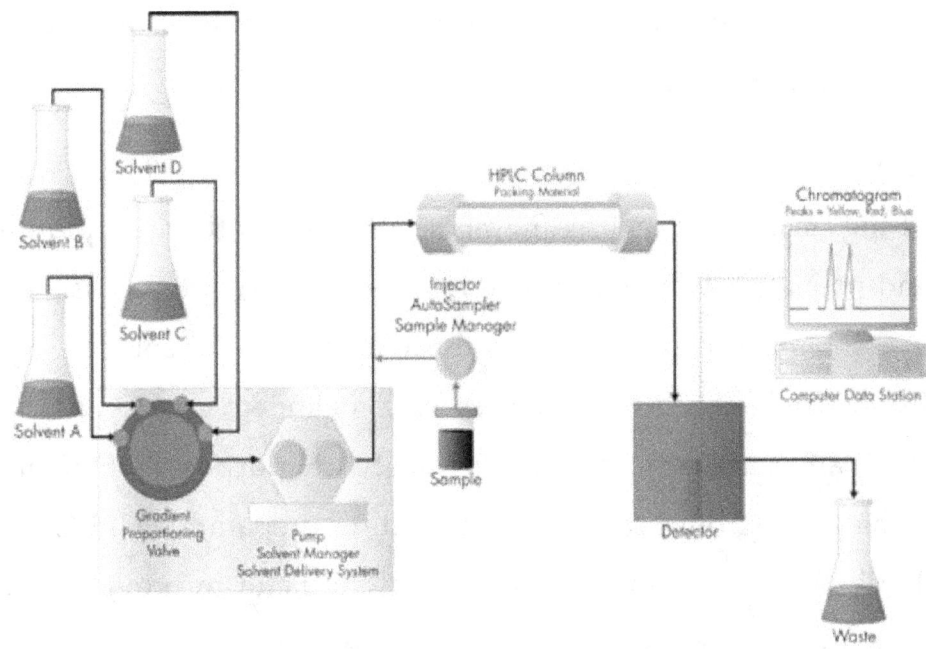

Figure – 3 Low-Pressure-Gradient System

Today's HPLC requires very special apparatus which includes the following.

1. Extremely precise gradient mixers.

2. HPLC high pressure pumps with very constant flow.

3. Unique high accuracy, low dispersion, HPLC sample valves.

4. Very high efficiency HPLC columns with inert packing materials.

5. High sensitivity low dispersion HPLC detectors.

6. High speed data acquisition systems.

7. Low dispersion connecting tubes for valve to column and column to detector.

HPLC Gradient mixtures

HPLC gradient mixers must provide a very precise control of solvent composition to maintain a reproducible gradient profile. This can be complicated in HPLC by the small elution volumes required by many systems. It is much more difficult to produce a constant gradient when mixing small volumes then when mixing large volumes. For low pressure systems this requires great precision in the operation of the miniature mixing General Introduction valves used and low dispersion flows throughout the mixer. For multi-pump high pressure systems it requires a very precise control of the flow rate while making very small changes of the flow rate.

HPLC Pumps

Because of the small particles used in modern HPLC, modern LC pumps need to operate reliably and precisely at pressures of 10,000 psi or at least 6,000 psi. To operate at these pressures and remain sensibly inert to the wide variety of solvents used HPLC pumps usually have sapphire pistons, stainless steel cylinders and return valves fitted with sapphire balls and stainless steel seats. For analytical purposes HPLC pumps should have flow rates that range from 0 to 10 ml/min., but for preparative HPLC, flow rates in excess of 100 ml/min may be required. It is extremely difficult to provide a very constant flow rate at very low flow rates. If 1% is considered acceptable then for 1ml/min a flow variation of less than 10µl/min is required. This level of constancy is required because most HPLC detectors are flow sensitive and errors in quantization will result from change in flow rate.

HPLC Sample Valves

Since sample valves come between the pump and the column it follows that HPLC sample valves must also tolerate pressures up to 10,000 psi. For analytical HPLC, the sample volume should be selectable from sub micro liter to a few micro liters, whereas in preparative HPLC the sample volume may be even greater than 10 ml. To maintain system efficiency the sample valve must be designed to have very low dispersion characteristics, this is true not only for flow dispersion but also for the less obvious problems of dispersion caused by sample adsorption/desorption on valve surfaces and diffusion of sample into and out of the mating surfaces between valve moving parts. It goes without saying that the valves must deliver a very constant sample size but this is usually attained by the use of a constant size sample loop.

HPLC Columns

HPLC columns are packed with very fine particles (usually a few microns in diameter). The very fine particles are required to attain the low dispersion that give the high plate counts expected of modern HPLC. Plate counts in excess of 25,000 plates per column are possible with modern columns, however, these very high efficiencies are very rarely found with real samples because of the dispersion associated with injection valves, detectors, data acquisition systems and the dispersion due to the higher molecular weight of real samples as opposed to the common test samples. Packing these small particles into the column is a difficult technical problem but even with good packing a great amount of care must be given to the column end fittings and the inlet and outlet connection to keep dispersion to a minimum. The main consideration with HPLC is the much wider variety of solvents and packing materials that can be utilized because of the much lower quantities of both which are required. In particular very expensive optically pure compounds can be used to make Chiral HPLC stationary phases and may even be used as (disposable) HPLC solvents.

HPLC Detectors [5-10]

UV/Vis spectrophotometers, including diode array detectors, are the most commonly employed detectors. Fluorescence spectrophotometers, differential refractometers, electrochemical detectors, mass spectrometers, light scattering detectors, radioactivity detectors or other special detectors may also be used. Detector consists of a flow-through cell mounted at the end of the column. A beam of UV radiation passes through the flow cell and into the detector. As compounds elute from the column, they pass through the cell and absorb the radiation, resulting in measurable energy level changes. Fixed (mercury lamp), variable (deuterium or high pressure xenon lamp), and multi-wavelength detectors are widely available. Modern variable wavelength detectors can be programmed to change wavelength while an analysis is in progress. Multi-wavelength detectors measure absorbance at two or more wavelengths simultaneously. In diode array multi-wavelength detectors, continuous radiation is passed through the sample cell, and then resolved into its constituent wavelengths, which are individually detected by the photodiode array. These detectors acquire absorbance data over the entire UV-visible range, thus providing the analyst with chromatograms at multiple, selectable wavelengths, spectra of the eluting peaks and also peak purity.

Differential refractometer detectors measure the difference between the refractive index of the mobile phase alone and that of the mobile phase containing chromatographic compounds as it emerges from the column. Refractive index detectors are used to detect non-UV absorbing compounds. Fluorometric detectors are sensitive to compounds that are inherently fluorescent or that can be converted to fluorescent derivatives either by chemical transformation of the compound or by coupling with fluorescent reagents at specific functional groups. Potentiometric, voltametric, or polarographic electrochemical detectors are useful for the quantitation of species that can be oxidized or reduced at a working electrode.

These detectors are selective, sensitive, and reliable, but require conducting mobile phases free of dissolved oxygen and reducible metal ions. Electrochemical detectors with carbon-paste electrodes may be used advantageously to measure nanogram quantities of easily oxidized compounds, notably phenols and catechols. In order to give an accurate chromatographic profile the detector sampling (cell) volume must be a small fraction of the solute elution volume. If the detector volume were larger than the elution volume then you would have peaks that appeared with flat tops as the whole peak would be resident in the detector at the same time. This means that as column volumes decrease and system efficiencies increase the volume of the detector cell volume must also decrease. It is odds for the requirement for detector to maintain high sensitivity as this is usually dependent on having a larger cell volume. Again, this requires the very careful design of modern detectors. Many types of detectors can use with HPLC system like UV-Visible or PDA (Photo Diode Array), RI (Refractive Index), Fluorescence, ECD (Electro Chemical Detector), ELSD (Evaporative Light Scattering detector) and many others hyphenated techniques like MS, MS/MS and NMR as well as evaporative IR.

HPLC Data acquisition

In HPLC data acquisition system the higher sampling rate needed for the rapidly eluting narrow peaks of the HPLC chromatogram. Although the theoretical number of samples needed for good quantization are actually quite small, for real systems a hundred samples or more per peak is recommended; thus, for a 4 sec wide peak, a rate of 25 samples per second may be required. The same data analysis and reporting software can be used as in ordinary LC.

Conclusion

HPLC is probably the most universal type of analytical procedure; its application areas include quality control, process control, forensic analysis, environmental monitoring and clinical testing. In addition HPLC also ranks as one of the most sensitive analytical procedures and is unique in that it easily copes with multi-component mixtures. It has achieved this position as a result of the constant evolution of the equipment used in LC to provide higher and higher efficiencies at faster and faster analysis times with a constant incorporation of new highly selective column packings.

3.1. Introduction to HPLC Methods of Analysis for Drugs [11-13]

Most of the drugs in single/multi component dosage forms can be analyzed by HPLC method because of the several advantages like rapidity, specificity, accuracy, precision and ease of automation in this method. HPLC method eliminates tedious extraction and isolation procedures. Some of the advantages are:

- ✓ Speed (analysis can be accomplished in 20 minutes or less).
- ✓ Greater sensitivity (various detectors can be employed).
- ✓ Improved resolution (wide variety of stationary phases).
- ✓ Reusable columns (expensive columns but can be used for many analysis).
- ✓ Ideal for the substances of low volatility.
- ✓ Easy sample recovery, handling and maintenance.
- ✓ Instrumentation tends itself to automation and quantitation.
- ✓ Precise and reproducible.

15

✓ Calculations are done by integrator itself.

✓ Suitable for preparative liquid chromatography on a much larger scale.

There are different modes of separation in HPLC. They are normal phase mode, reversed phase mode, reverse phase ion pair chromatography, affinity chromatography and size exclusion chromatography (gel permeation and gel filtration chromatography). In the normal phase mode, the stationary phase is polar and the mobile phase is non-polar in nature. In this technique, non-polar compounds travel faster and are eluted first. This is because of the lower affinity between the non-polar compounds and the stationary phase. Polar compounds are retained for longer times because of their higher affinity with the stationary phase. These compounds, therefore take more times to elute.

Normal phase mode of separation is therefore, not generally used for pharmaceutical applications because most of the drug molecules are polar in nature and hence takes longer time to elute. Reversed phase mode is the most popular mode for analytical and preparative separations of compound of interest in chemical, biological, pharmaceutical, food and biomedical sciences. In this mode, the stationary phase is non-polar hydrophobic packing with octyl or octa decyl functional group bonded to silica gel and the mobile phase is polar solvent. An aqueous mobile phase allows the use of secondary solute chemical equilibrium (such as ionization control, ion suppression, ion pairing and complexation) to control retention and selectivity. The polar compound gets eluted first in this mode and nonpolar compounds are retained for longer time. As most of the drugs and pharmaceuticals are polar in nature, they are not retained for longer times and hence elute faster. The different columns used are Octa Decyl Silane (ODS) or C18, C8, C4, etc. (in the order of increasing polarity of the stationary phase).

In ion exchange chromatography, the stationary phase contains ionic groups like NR_3^+ or SO_3^{-2}, which interact with the ionic groups of the sample molecules. This is suitable for the separation of charged molecules only. Changing the pH and salt concentration can modulate the retention. Ion pair chromatography may be used for the separation of ionic compounds and this method can also substitute for ion exchange chromatography. Strong acidic and basic compounds may be separated by reversed phase mode by forming ion pairs (columbic association species formed between two ions of opposite electric charge) with suitable counter ions. This technique is referred to as reversed phase ion pair chromatography or soap chromatography.

Affinity chromatography uses highly specific biochemical interactions for separation. The stationary phase contains specific groups of molecules which can absorb the sample if certain steric and charge related conditions are satisfied. This technique can be used to isolate proteins, enzymes as well as antibodies from complex mixtures.

Size exclusion chromatography separates molecules according to their molecular mass. Largest molecules are eluted first and the smallest molecules last. This method is generally used when a mixture contains compounds with a molecular mass difference of at least 10%. This mode can be further subdivided into gel permeation chromatography (with organic solvents) and gel filtration chromatography (with aqueous solvents).

Method Development and Design of Separation Method

Methods for analyzing drugs in single or multi component dosage forms can be developed, provided one has knowledge about the nature of the sample, namely, its molecular weight, polarity, ionic character and the solubility parameter. An exact recipe for HPLC, however, cannot be provided because method development involves considerable trial and error procedures. The most difficult problem usually is where to start, what type of column is worth trying with what kind of mobile phase. In general one begins with reversed phase chromatography, when the compounds are hydrophilic in nature with many polar groups and are water soluble.

The organic phase concentration required for the mobile phase can be estimated by gradient elution method. For aqueous sample mixtures, the best way to start is with gradient reversed phase chromatography. Gradient can be started with 5-10 % organic phase in the mobile phase and the organic phase concentration (methanol or acetonitrile)

17

can be increased up to 100 % within 30-45 min. Separation can then be optimized by changing the initial mobile phase composition and the slope of the gradient according to the chromatogram obtained from the preliminary run. The initial mobile phase composition can be estimated on the basis of where the compounds of interest were eluted, namely at what mobile phase composition.

Changing the polarity of mobile phase can alter elution of drug molecules. The elution strength of a mobile phase depends upon its polarity, the stronger the polarity, higher is the elution. Ionic samples (acidic or basic) can be separated, if they are present in undissociated form. Dissociation of ionic samples may be suppressed by the proper selection of pH.

The pH of the mobile phase has to be selected in such a way that the compounds are not ionized. If the retention times are too short, the decrease of the organic phase concentration in the mobile phase can be in steps of 5%. If the retention times are too long, an increase of the organic phase concentration is needed. In UV detection, good analytical results are obtained only when the wavelength I selected carefully. This requires knowledge of the UV spectra of the individual components present in the sample. If analyte standards are available, their UV spectra can be measured prior to HPLC method development.

The molar absorbance at the detection wavelength is also an important parameter. When peaks are not detected in the chromatograms, it is possible that the sample quantity is not enough for the detection. An injection of volume of 20 µL from a solution of 1 mg/mL concentration normally provides good signals for UV active compounds around 220 nm. Even if the compounds exhibit higher λmax, they absorb strongly at lower wavelength.

It is not always necessary to detect compounds at their maximum absorbance. It is, however, advantageous to avoid the detection at the sloppy part of the UV spectrum for precise quantitation. When acceptable peaks are detected on the chromatogram, the investigation of the peak shapes can help further method development. The addition of peak modifiers to the mobile phase can affect the separation of ionic samples. For examples, the retention of the basic compounds can be influenced by the addition of small amounts of triethylamine (a peak modifier) to the mobile phase. Similarly for acidic compounds small amounts of acids such as acetic acid can be used. This can lead to useful changes in selectivity. When tailing or fronting is observed, it means that the mobile phase is not totally compatible with the solutes. In most case the pH is not properly selected and hence partial dissociation or protonation takes place. When the peak shape does not improve by lower (1-2) or higher (8-9) pH, then ion-pair chromatography can be used. For acidic compounds, cationic ion pair molecules at higher pH and for basic compounds, anionic ion-pair molecules at lower pH can be used. For amphoteric solutes or a mixture of acidic and basic compounds, ion-pair chromatography is the method of choice. The low solubility of the sample in the mobile phase can also cause bad peak shapes. It is always advisable to use the same solvents for the preparation of sample solution as the mobile phase to avoid precipitation of the compounds in the column or injector.

Optimization can be started only after a reasonable chromatogram has been obtained. A reasonable chromatogram means that more or less symmetrical peaks on the chromatogram detect all the compounds. By sight change of the mobile phase composition, the position of the peaks can be predicted within the range of investigated changes. An optimized chromatogram is the one in which all the peaks are symmetrical and are well separated in less run time.

The peak resolution can be increased by using a more efficient column (column with higher theoretical plate number, N) which can be achieved by using a column of smaller particle size, or a longer column. These factors, however, will increase the analysis time. Flow rate does not influence resolution, but it has a strong effect on the analysis time.

Unfortunately, theoretical predictions of mobile phase and stationary phase interactions with a given set of sample components are not always accurate, but they do help to narrow down the choices for method development. The separation scientist must usually perform a series of trial and error experiments with different mobile phase compositions until a satisfactory separation is achieved.

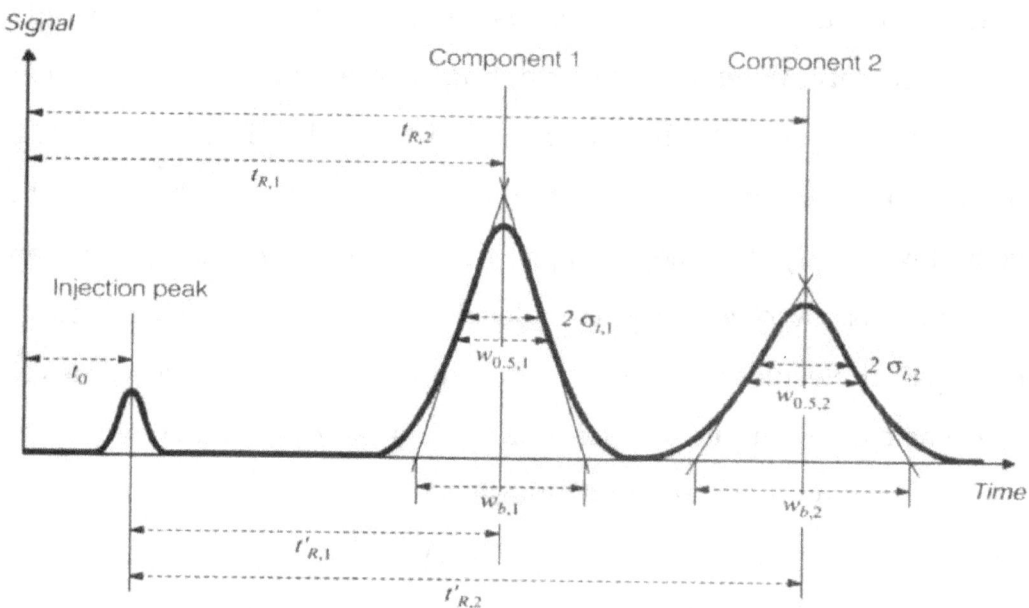

The parameters that are affected by the changes in chromatographic conditions are:

1. Resolution (Rs).

2. Capacity factor (k').

3. Selectivity (α).

4. Plate number (N).

5. Asymmetry factor (T).

1. Resolution (Rs): Resolution is the parameter describing the separation power of the complete chromatographic system relative to the particular components of the mixture.

$$R = \frac{2(tR,2 - tR,1)}{Wb,1 + Wb,2} = \frac{1.177(tR,2 - tR,1)}{W0.5,1 + W0.5,2}$$

If the peak base widths $w_{b.1}$ and $w_{b.2}$ are approximately the same, the resolution R signifies the number of times the peak width w_b can be fitted into the distance between the peak maxima. At a resolution of R=0.5, two maxima can still be perceived separately. For quantitative analysis, a resolution of up to R=1.5 is desirable; greater values of the resolution lead only to unnecessarily long analysis times. The resolution R is dependent on the parameters k_2' (capacity factor of the later eluted substance), selectivity α and plate number N of the column:

$$R = \frac{\sqrt{N}}{4} \times \frac{\alpha - 1}{\alpha} \times \frac{k2'}{1 + k2'}$$

2. Capacity factor (k'): The retention time t_R is the qualitative information of a chromatogram. It is constant for a given component provided the chromatographic conditions remain unchanged (column, mobile phase, temperature etc.) for the characterization of substance, it is more convenient to quote the capacity factor k' since, in contrast to the retention times, this is dependent neither on the flow of the eluent nor on the column length:

$$k' = \frac{tR'}{t0} = \frac{tR - t0}{t0} = \frac{tR}{t0} - 1$$

3. Selectivity (α): The selectivity (or separation factor) is a measure of relative retention of two components in a mixture. Selectivity is the ratio of the capacity factors of both peaks, and the ratio of its adjusted retention times. Selectivity represents the separation power of particular adsorbent to the mixture of these particular components. This parameter is independent of the column efficiency; it only depends on the nature of the components, eluent type, eluent composition and adsorbent surface chemistry. In general, if the selectivity of two components is equal to 1, then there is no way to separate them by improving the column efficiency. The ideal value of α is 2. It can be calculated by using formula,

$$\alpha = \frac{k2'}{k1'} = \frac{tR,2 - t0}{tR.1 - t0} \qquad (k2' > k1')$$

4. Plate number (N): An additional useful quantity to characterize a separation system is the plate number N (number of theoretical plates). A theoretical plate is defined as that zone of separation system within which a thermodynamic equilibrium is established between the mean concentration of a component in the stationary phase and its mean concentration in the mobile phase. Efficiency is calculated by using the formula,

$$N = 16 \, (t_R / w_b)^2$$

Where, t_R is the retention time.

w_b is the peak width.

5. Asymmetry factor (T): The elution of chromatographic signals as Gaussian peaks is often not achieved in practice. An asymmetric peak shape, known as tailing, is often found. The peak asymmetry is quantified by the asymmetry factor (tailing factor) T with a and b being determined at 10 % peak height:

$$T = \frac{b}{a}$$

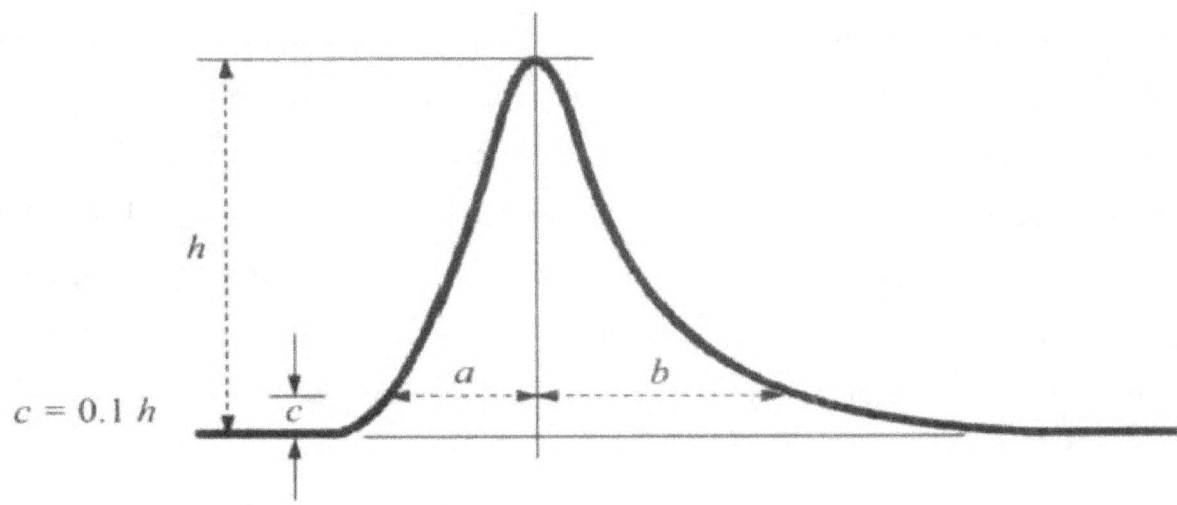

For the trouble-free evaluation of the area of a peak, T must be < 2.5, above this, the end of the peak can be recognized only with difficulty. For a well-packed column, an asymmetry factor of 0.9 to 1.1 should be achievable.

3.2. Separation Mode of HPLC

Various modes of HPLC utilized to separate compounds are classified as follows:

a) Adsorption chromatography

b) Normal-phase chromatography

c) Reversed-phase chromatography

d) Ion-pair chromatography

e) Ion-exchange chromatography

f) Size exclusion chromatography

a) Adsorption chromatography

Adsorption chromatography uses polar stationary phases with relatively non-polar mobile phases. Separations in adsorption chromatography result to a great extent from the interaction of sample polar functional groups with discrete adsorption sites on the stationary phase. Adsorption chromatography is usually considered appropriate for the separation of nonionic molecules that are soluble in organic solvents.

b) Normal-phase chromatography

In HPLC, if stationary phase is more polar than the mobile phase, it is termed as normal phase liquid chromatography. Polar bonded phases that have a diol, cyano, diethylamino, amino, or diamino functional groups are used as stationary phase in normal-phase chromatography. Due to lower affinity of non-polar compounds to the stationary phases used, non-polar compounds are elute first while polar compounds are retained for longer time. Normal-phase chromatography is widely applied for chiral separations.

c) Reversed-phase chromatography

In HPLC, if stationary phase is less polar than the mobile phase, it is termed as reversed phase liquid chromatography. In this technique, C18, C8, Phenyl, and cyano-propyl functional groups that chemically bonded to micro porous silica particles are used as stationary phase. Retention in reversed phase chromatography occurs by nonspecific hydrophobic interactions of the solute with stationary phase. The ubiquitous application of reversed-phase chromatography arise from the fact that practically all organic molecules have hydrophobic regions in their structures and effectively interact with the stationary phase. It is estimated that over 65% (possibly as high as 90%) of all HPLC separations are executed in the reversed-phase mode. The rationale for this includes the simplicity, versatility, and scope of the reversed-phase method [14].

d) Ion-pair chromatography

Ionic or partially ionic compounds can be chromatographed on reversed phase columns by using ion-pairing reagents. These reagents are typically long chain alkyl anions or cations that, when used in dilute concentrations, can increase the retention of analyte ions. C-5 to C-10 alkyl sulfonates are commonly used for cationic compounds while C-5 to C-8 alkyl ammonium salts are generally used in the cases of anionic solutes.

e) Ion-exchange chromatography

Ion-exchange chromatography is an adaptable technique used primarily for the separation of ionic or easily ionizable species. The stationary phase is characterized by the presence of charged centers having exchangeable counter ions. Both anions and cations can be separated by choosing the suitable ion-exchange medium. Ion-exchange chromatography employs the dynamic interactions between charged solute ions and stationary phases that have oppositely charged groups.

f) Size exclusion chromatography

Size exclusion chromatography separates molecules according to their molecular mass. In Size exclusion chromatography, column is filled with material having precisely controlled pore sizes and the sample is simply screened or filtered according to its solvated molecular size. Largest molecules are eluted first and the smallest molecules last. This method is generally used when a mixture contains compounds with a molecular mass difference of at least 10%. This mode can be further subdivided into gel permeation chromatography (with organic solvents) and gel filtration chromatography (with aqueous solvents).

4. Analytical method validation

The developed analytical procedure used to measure the quality of pharmaceutical products. It is necessary to assure that the performance characteristics of the developed analytical procedure meet the requirements for the intended analytical application. The procedure which provides assurance for the same by the means of laboratory studies is defined as method validation. Method validation is the process of demonstrating that analytical procedures are suitable for their intended use and that they support the identity, strength and quality, for the quantification of the drug substances and drug products. Method validation has received considerable attention in the literature and from industrial committees and regulatory agencies. The U.S. FDA cGMP [15] states for validation for the test methods employed by the firm. The U.S. FDA has also proposed industry guidance for Analytical Procedures and Methods Validation [16]. ISO/IEC 17025 includes a chapter on the validation of methods [17] with list validation parameters. The ICH [18] has developed a consensus text on the validation of analytical procedures. ICH also developed guidance with detailed methodology [19]. The U.S. EPA prepared guidance for method's development and validation for the Resource Conservation and Recovery Act (RCRA) [20].

The AOAC, the EPA and other scientific organizations provide methods that are validated through multi-laboratory studies. The USP has published specific guidelines for method validation for compound evaluation [21]. The WHO published validation guidelines under the title, 'Validation of analytical procedures used in the examination of pharmaceutical materials' in the 32nd report of the WHO expert committee on 'specifications for pharmaceutical preparations'. Representatives of the pharmaceutical and chemical industry have published papers on the validation of analytical methods.

Hokanson [22, 23] applied the life cycle approach, developed for computerized systems, to the validation and revalidation of methods. Green [24] gave a practical guide for analytical method validation, with a description of a set of minimum requirements for a method. Wegscheider [25] has published procedures for method validation with a special focus on calibration, recovery experiments, method comparison and investigation of ruggedness. Seno et al. [26] have described how analytical methods are validated in a Japanese QC laboratory. The AOAC [27] has developed a Peer-Verified Methods validation program with detailed guidelines on exactly which parameters should be validated. Winslow and Meyer [28] recommend the definition and application of a master plan for validating analytical methods. J. Breaux and colleagues have published a study on analytical methods development and validation [29].

4.1.   Strategy for the Validation of Methods

Method development and validation are an iterative process. The influence of operating parameters on the performance of the method can be assessed at the validation stage which was not done during development/optimization stage of the method. The most significant point raised for validation is that the validity of a method can be demonstrated only through laboratory studies. It is not sufficient to simply review historical results; instead, laboratory studies must be conducted which are intended to validate the specific method, and those studies should be pre-planned and described in a suitable documentation. This documentation should clearly indicate the method's intended use and principles of operation, as well as the validation parameters to be studied, and a rationale for why this method and these parameters were chosen. It also must include pre-defined acceptance criteria and a description of the analytical procedure.

4.2.   Parameters for Method Validation

The parameters for method validation have been defined in different working groups of national and international committees and are described in the literature. An attempt at harmonization was made for pharmaceutical applications through the ICH [18, 19]. The defined validation parameters by the ICH and other regulatory bodies are summarized as under:

a) Specificity study

b) Linearity and range study

c) Limit of detection and Limit of quantitation study

d) Precision study

e) Accuracy study

f) Robustness study

g) Solution stability study

h) System suitability

A brief introduction of above parameters is as below:

a) Specificity study

Specificity of an analytical method is its ability to measure accurately an analyte in the presence of interference, such as synthetic precursors, excipients, enantiomers and known (or likely) degradation products that may be expected to be present in the sample matrix. The term specificity is also referring to selectivity when a number of chemical entities that may or may not be distinguished from each other.

Specificity study should also assess interferences that may be caused by the matrix, e.g., urine, blood, soil, water or food. Optimized sample preparation can eliminate most of the matrix components, e.g. placebo. The absence of matrix interferences for a quantitative method should be demonstrated by the analysis of control matrix in specificity. An investigation of specificity should be conducted during the validation of identification tests, the determination of impurities and the assay. In order to check the interference of degradation products, analyte is forcibly subject to chemical (acid, alkali and oxidative) and physical (thermal and photolytic) degradation, known as stress application. In each stress application, peak purity of the analyte peak is also evaluated [30].

b) Linearity and range study

The linearity of an analytical method is its ability to elicit test results that are directly proportional to the concentration (amount) of analyte in samples within a given range. Linearity may be demonstrated directly on the test substance (by dilution of a standard stock solution) and/or by using separate amounts of synthetic mixtures of the test product components, using the proposed procedure.

Linearity is determined by replicate injections of 5 or more concentrations level within the range of 40–160 %. The response should be directly proportional to the concentrations of the analytes or proportional by means of a well-defined mathematical calculation. Linearity is evaluated graphically by plotting a graph of the relative responses on the y-axis and the corresponding concentrations on the x-axis. A linear regression equation is applied to the results to evaluate correlation coefficient. In addition, y-intercept, slope of the regression line and residual sum of squares should also calculate.

The range of an analytical method is the interval between the upper and lower concentrations (amounts) of analyte in the sample (including these concentrations) for which it has been demonstrated that the analytical procedure has a suitable level of precision, accuracy and linearity. The range is normally expressed in the same units as the test results (e.g., percentage, parts per million) obtained by the analytical method.

c) Limit of detection (LOD) and Limit of quantitation (LOQ) study

The detection limit of an analytical method is the lowest amount of analyte in a sample which can be detected but not necessarily quantitated as an exact value. In chromatography, the detection limit is the injected amount that results in a peak with a height at least two or three times as high as the baseline noise level. Besides this signal/noise method, LOD can be measured by another three different methods; (i) visual inspection (ii) standard deviation of the blank response (iii) standard deviation of the response based on the slope of the calibration curve.

The quantitation limit of an analytical method is the lowest amount of analyte in a sample which can be quantitated with suitable precision and accuracy. In chromatography, the quantitation limit is the minimum injected amount that produces quantitative measurements in the target matrix with acceptable precision, typically requiring peak heights 10 to 20 times higher than the baseline noise. Beside this signal/noise method, LOQ can be measured by another three different methods; (i) visual inspection (ii) standard deviation of the response (iii) standard deviation of the response based on the slope of the calibration curve.

d) Precision study

The precision of an analytical method expresses the closeness of agreement (degree of scatter) between a series of measurements obtained from multiple sampling of the homogeneous sample under the prescribed conditions.

The measurement of precision of an analytical method is performed on replicate standard preparations and replicate sample preparations. The results for the same are usually expressed as the variance, standard deviation or confidence level of a series of measurements. Precision is performed by means of repeatability, reproducibility and intermediate precision (ruggedness)

Repeatability: Repeatability expresses the precision under the same operating conditions over a short interval of time. Repeatability is also termed intra-assay precision.

Reproducibility: Reproducibility expresses the precision between laboratories. The reproducibility of an analytical method is determined by analyzing aliquots from same homogeneous lots.

Intermediate Precision: Intermediate precision expresses within-laboratories variations; different days, different analysts, different equipment, etc. The objective of intermediate precision validation is to verify that in the same laboratory the method will provide the same results once the development phase is over. The objective is also extent to verify that the method will provide the same results in different laboratories (ruggedness).

e) Accuracy study

The accuracy of the analytical method is the closeness of agreement between the value which is accepted either as a conventional true value or an accepted reference value, and the value found. The accuracy of an analytical method is the extent to which test results generated by the method and the true value agree. The true value for accuracy assessment can be assessed by analyzing a sample with known concentrations (e.g., a control sample or certified reference material) and comparing the measured value with the true value as supplied with the material. If certified reference materials or control samples are not available, a blank sample matrix of interest can be spiked with a known concentration by weight or volume. After extraction of the analyte from the matrix and injection into the analytical instrument, its recovery can be determined by comparing the response of the extract with the response of the reference material dissolved in a pure solvent (without matrix).

The ICH document on validation methodology recommends accuracy to be assessed using a minimum of nine determinations over a minimum of three concentration levels covering the specified range (e.g., three concentrations/three replicates each). Accuracy should be reported as percent recovery by the assay of known added amount of analyte in the sample or as the difference between the mean and the accepted true value.

f) Robustness study

The robustness of an analytical method is a measure of its capacity to remain unaffected by small, but deliberate, variations in method parameters and provides an indication of its reliability during normal usage. Robustness tests examine the effect that operational parameters have on the analysis results. For the determination of a method's robustness, method parameters like pH, flow rate, column temperature, column lot or mobile phase composition, are varied within a realistic range, and the quantitative influence of the variables is determined. If the influence of the parameter is within a previously specified tolerance, the parameter is said to be within the method's robustness range.

g) Solution stability study

Many solutes readily decompose prior to chromatographic investigations, for example, during the preparation of the sample solutions, extraction, cleanup, phase transfer or storage of prepared vials (in refrigerators or in an automatic sampler). Under these circumstances, method validation should investigate the stability of the analytes and standards in solution form (in analytical preparations). The standard and test preparations are stored up to specified period at specified temperature and its stability is evaluated by comparing solution preparations at different time intervals to that of initial.

h) System suitability study

In addition, prior to the start of laboratory studies to demonstrate method validity, some type of system suitability must be done to demonstrate that the analytical system is performing properly. System suitability should be determined by replicate analysis of the standard or reference solution. System suitability is considered appropriate when the RSD, theoretical plates, tailing factor and resolution parameters calculated on the results obtained at different time intervals, does not exceed more than of specified limit of the corresponding value of the system precision.

4.3.   Prerequisites of Validation

Prior to start method validation, validation aim should be a well-planned according to scientific soundness and completeness with pre-defined acceptance criteria. Because the type of analysis and the other information of a sample have so much influence on the validation, the objective and scope of the method should always be defined as the first step of any method validation. For an efficient validation process, it is most importance to specify the right validation parameters.

Subsequent to the execution of the validation, results, conclusions and deviations should present in report. Provided the pre-defined acceptance criteria are met, and the deviations

31

(if any) do not affect the scientific interpretation of the data, then the developed analytical method can be considered as valid.

5. Objectives of the Work

The objectives of the book were achieved in the following three phases:

(a) Development of new HPLC method for the determination of APIs in Pharmaceutical products.

(b) System suitability testing of newly developed HPLC methods.

(c) Validation of the developed method.

Phase I:

5.1. Introduction of Drug

Platelet aggregation and thrombus formation play a critical role in the initiation and development of key complications of acute coronary syndromes (ACSs). Antiplatelet therapy and antithrombotic therapy have been demonstrated to favorably modify clinical outcome, and recent trials of revascularization in ACSs have demonstrated a reduction in the frequency of major cardiac events [1-13]. Antiplatelet and antithrombin therapy can have synergistic actions that reduce the risk of spontaneous or revascularization, especially percutaneous coronary intervention (PCI)-related events. Yet, all effective antithrombotic agents also increase the risk of bleeding, especially bleeding that results from vascular accessor associated with surgery, including coronary artery bypass grafting (CABG). The Clopidogrel in Unstable angina to prevent recurrent ischemic Events (CURE) trial demonstrated that the combination of clopidogrel and aspirin was superior to aspirin alone for patients hospitalized with non-ST-elevation ACSs.

5.2. Clopidogrel bisulphate

(1) Method objectives and understand the chemistry of Clopidogrel bisulphate

Clopidogrel bisulfate, chemically it is [S- (a)(2-chlorophenyl)-6,7- dihydrothieno (3,2-C) pyridine-5 (4H) acetic acid methyl ester sulphate] (Figure 2). The empirical formula of clopidogrel bisulfate is $C_{16}H_{16}ClNO_2S \cdot H_2SO_4$ and its molecular weight is 419.9 g/mol [14]. It is a white to off-white powder. It is practically insoluble in water at neutral pH but freely soluble at pH 1. It also dissolves freely in methanol, dissolves sparingly in methylene chloride and is practically insoluble in ethyl ether. It has a specific optical rotation of about +56°. The structural formula is as follows:

Chemical Structure of clopidogrel

Pharmacology

Mechanism of Action: Clopidogrel is an inhibitor of platelet aggregation. A variety of drugs that inhibit platelet function have been shown to decrease morbid events in people with established cardiovascular atherosclerotic disease as evidenced by stroke or transient ischemic attacks, myocardial infarction, unstable angina or the need for vascular by-pass or angioplasty. This indicates that platelets participate in the initiation and/or evolution of these events and that inhibiting them can reduce the event rate.

Clopidogrel selectively inhibits the binding of adenosine diphosphate (ADP) to its platelet receptor and the subsequent ADP-mediated activation of the glycoprotein GPIIb/IIIa complex, thereby inhibiting platelet aggregation. Biotransformation of clopidogrel is necessary to produce inhibition of platelet aggregation, but an active metabolite responsible for the activity of the drug has not been isolated. Clopidogrel also inhibits platelet aggregation induced by agonists other than ADP by blocking the amplification of platelet activation by released ADP. Clopidogrel does not inhibit phosphodiesterase activity.

Clopidogrel acts by irreversibly modifying the platelet ADP receptor. Consequently, platelets exposed to clopidogrel are affected for the remainder of their lifespan. Dose dependent inhibition of platelet aggregation can be seen 2 hours after single oral doses of clopidogrel bisulfate. Repeated doses of 75 mg clopidogrel bisulfate per day inhibit ADP induced platelet aggregation on the first day and inhibition reaches steady state between day 3 and day 7. At steady state, the average inhibition level observed with a dose of 75 mg clopidogrel bisulfate per day was between 40% and 60%. Platelet aggregation and bleeding time gradually return to baseline values after treatment is discontinued, generally in about 5 days.

Pharmacokinetics

After repeated 75 mg oral doses of clopidogrel (base), plasma concentrations of the parent compound, which has no platelet inhibiting effect, are very low and are generally below the quantification limit (0.00025 mg/L) beyond 2 hours after dosing. Clopidogrel is extensively metabolized by the liver. The main circulating metabolite is the carboxylic acid derivative, and it too has no effect on platelet aggregation. It represents about 85% of the circulating drug related compounds in plasma.

Following an oral dose of 14C-labeled clopidogrel in humans, approximately 50% was excreted in the urine and approximately 46% in the feces in the 5 days after dosing. The elimination half-life of the main circulating metabolite was 8 hours after single and repeated administration. Covalent binding to platelets accounted for 2% of radiolabel with a half-life of 11 days.

Effect of food: Administration of clopidogrel bisulfate with meals did not significantly modify the bioavailability of clopidogrel as assessed by the pharmacokinetics of the main circulating metabolite.

Absorption and distribution: Clopidogrel is rapidly absorbed after oral administration of repeated doses of 75 mg clopidogrel (base), with peak plasma levels (3 mg/L) of the main circulating metabolite occurring approximately 1 hour after dosing. The pharmacokinetics of the main circulating metabolite are linear (plasma concentrations increased in proportion to dose) in the dose range of 50 to 150 mg of clopidogrel. Absorption is at least 50% based on urinary excretion of clopidogrel related metabolites.

Clopidogrel and the main circulating metabolite bind reversibly in vitro to human plasma proteins (98% and 94%, respectively). The binding is nonsaturable *in vitro* up to a concentration of 100 µg/ml.

Metabolism and elimination: In vitro and *in vivo*, clopidogrel undergoes rapid hydrolysis into its carboxylic acid derivative. In plasma and urine, the glucuronide of the carboxylic acid derivative is also observed.

(2) Initial HPLC Condition

Mobile phase: Buffer-ACN (60: 40, v/v) Buffer: 0.3% orthophosphoric acid

Column: Phenomenex, C8 (250 mm x 4.6 mm i.d., 5µ particle size)

Flow rate: 1 ml/min

Wavelength: 226 nm

Injection volume: 20 µl

Diluent: Mobile phase

(3) Sample Preparation

a. Standard Preparation

Standard solution containing aspirin (0.075 mg/ml) and clopidogrel (0.0375 mg/ml) were prepared by dissolving 37.5 mg aspirin and 24.46 mg clopidogrel bisulphate (equivalent to 18.5 mg clopidogrel) in 100 ml volumetric flask by mobile phase (stock standard solution). Pipette out 10 ml stock solution into 50 ml volumetric flask and dilute up to mark with mobile phase (standard solution).

b. Test Preparation

Twenty tablets were weighed and the average tablet weight was determined. Tablets were crushed by mortar and pastel. Tablet powder was weighed equivalent to five times of average weight and transfer in to 200 ml volumetric flask. About 170 ml mobile phase was added and sonicated for of 30 min time interval with intermittent shaking. Content was brought back to room temperature and dilute to volume with mobile phase (stock solution). The stock solution was filtered through 0.45 µm nylon syringe filter. Pipette out 2 ml filtered stock solution in to 100 ml volumetric flask and diluted with mobile phase (test solution). The concentration obtain was 0.075 mg/ml of aspirin and 0.0375 mg/ml of clopidogrel.

(4) Literature Survey

(1) H. Agrawal, N. Kaul, A. R. Paradkar, K. R. Mahadik have developed and validated stability indicating high-performance thin layer chromatographic method of analysis of clopidogrel bisulphate both as a bulk drug and in formulations. The method employed TLC aluminium plates pre-coated with silica gel 60F254 as the stationary phase. The solvent system consisted of carbon tetrachloride-chloroform-acetone (6:4:0.15, v/v/v). Clopidogrel bisulphate was subjected to acid and alkali hydrolysis, oxidation, photo degradation and dry heat treatment. The method could be employed as a stability indicating one.

(2) **E. Souri, H. Jalalizadeh, A. Kebriaee-Zadeh, M. Shekarchi, A. Dalvandi** have developed reproducible method for determination of carboxylic acid metabolite of clopidogrel in human plasma. After liquid-liquid extraction in acidic medium with chloroform, samples were quantified on a C8, 5 mm column using a mixture of 30 mm dipotassium hydrogen phosphate (pH 3)- tetra hydro furan-acetonitrile (79:2:19, v/v/v) as mobile phase with UV detection at 220 nm. The flow rate was set at 0.9 ml/min. Ticlopidine was used as internal standard and the total run time of analysis was about 12 min. The method was used to study the pharmacokinetics of clopidogrel.

(3) **S. S. Singh, K. Sharma, D. Barot, P. R. Mohan, V. B. Lohray** have developed a high-performance liquid chromatographic method for the estimation of carboxylic acid metabolite of clopidogrel bisulfate in rat plasma using atorvastatin as internal standard. Plasma samples were extracted with a mixture of ethyl acetate-dichloro methane (80:20, v/v) followed by subsequent reconstitution in a mixture of water: methanol: acetonitrile (40:40:20, v/v/v). The chromatographic separation was achieved with gradient elution on Kromasil ODS (250 mm x 4.6 mm i.d., 5 μm particle size) analytical column maintained at 30 °C. Carboxylic acid metabolite of clopidogrel as well as the internal standard were detected at a wavelength of 220 nm. The method was applied to the pharmacokinetic study of the two different polymorphs of clopidogrel bisulfate in wister rat.

(4) **A. Robinson, J. Hillis, C. Neal, A. C. Leary** have developed and validated LCMS/MS bio-analytical method for the determination of unchanged clopidogrel in human plasma. Analysis was performed using a C 8 column (temperature controlled to 50 °C) by gradient elution at a flow rate of 0.9 ml/min over a 3 min run time. Detection was achieved using a Sciex API 4000, triple quadrupole mass spectrometer, in positive turboionspray (electrospray) ionization mode. This validated method was used to support a pharmacokinetic study in healthy volunteers.

(5) **R. V. Nirogi, V. N. Kandikere, M. Shukla, K. Mudigonda, S. Maurya, R. Boosi** have reported high-performance liquid chromatography/positive electrospray ionization tandem mass spectrometry method for the quantification of clopidogrel in human plasma. The analytes were separated using an isocratic mobile phase on a reversed-phase column and analyzed by mass spectrometry in the multiple reaction monitoring mode. The validated method has been used to analyze human plasma samples for application in pharmacokinetic, bioavailability or bioequivalence studies.

(6) **H. Ksycinska, P. Rudzki, M. Bukowska-Kiliszek** have reported a method for determination of clopidogrel metabolite (SR26334) in human plasma. Samples were quantified using reversed phase high performance liquid chromatography with mass detection. The determination was performed on a Luna C18, (75 mm x 4.6 mm i.d., 3 μm particle size) column with an acetonitrile-water-formic acid mixture (60:40:0.1, v/v/v) as a mobile phase. The flow rate was set at 0.2 ml/min. The method has been used to study clopidogrel metabolite pharmacokinetics in healthy volunteers.

(7) **A. Mitakos, I. Panderi** have documented a reversed phase HPLC method for the determination of clopidogrel in pharmaceutical dosage forms. The determination was performed on a semi-micro column BDS

C8 (250 mm x 2.1 mm id., 5 µm particle size). The mobile phase consisted of a mixture of 0.010 Msodium dihydrogen phosphate (pH 3.0)-acetonitrile (35:65, v/v), pumped at a flow rate 0.3 ml/min. The UV detector was operated at 235 nm. The method was applied in the quality control of commercial tablets and content uniformity test.

(8) P. Lagorce, Y. Perez, J. Ortiz, J. Necciari, F. Bressolle haved eveloped GCMS method for the analysis of the carboxylic acid metabolite clopidogrel in plasma and serum. The analytical procedure involves a robotic liquid-liquid extraction with diethyl ether followed by a solid-liquid extraction on C18 cartridges. The derivatization process was performed using n-ethyl di-isopropylethylamine and alpha-bromo-2,3,4,5,6-pentafluoro toluene. The method use for the pharmacokinetic studies

5.3. Aceclofenac

(1) Method objectives and understand the chemistry of Aceclofenac.

Aceclofenac, 2-[(2,6-dichlorophenyl) amino]-phenylacetoxyacetic acid. The molecular formula of aceclofenac is $C_{16}H_{13}Cl_2NO_4$ and its molecular weight is 354.19 g/mole. It is a white or almost white powder. It is practically insoluble in water, freely soluble in acetone and soluble in alcohol.

Chemical structure of aceclofenac

That shows analgesic properties and good tolerability profile in a variety of painful conditions [5, 6]. It is used in the treatment of rheumatic disorders and soft tissue injuries. Aceclofenac inhibits the cyclooxygenase enzyme and thus exerts its anti-inflammatory activity by inhibition of prostaglandin synthesis. This effect seems to be correlated to the appearance of acute protocolitis associated with non-steroidal anti-inflammatory drug therapy.

Pharmacology

Mechanism of Action: The mode of action of aceclofenac is largely based on the inhibition of prostaglandin synthesis. Aceclofenac is potent inhibiter of the enzyme cyclo-oxygenase, which is involved in the production of prostaglandins. Aceclofenac has been shown to exert effects on a variety of mediators of inflammation. The drug inhibits synthesis of the inflammatory cytokines interleukin (IL)-1â and tumour necrosis factor and inhibits prostaglandin E2 (PGE2) production. Effects on cell adhesion molecules from neutrophils have also been noted. *In vitro* data indicate inhibition of cyclooxygenase (COX)-1 and 2 by aceclofenac in whole blood assays, with selectivity of COX-2 being evident.

In contrast to some other NSAIDs, aceclofenac has shown stimulatory effects on artilage matrix synthesis, which may be linked to the ability of the drug to inhibit IL-1â activity.

In vitro data indicates stimulation by the drug of synthesis of glycosaminoglycan in osteoarthritic cartilage. There is also evidence that aceclofenac stimulates the synthesis of IL-1 receptor antagonist in human articular chondrocytes subjected to inflammatory stimuli and that 4'-hydroxyaceclofenac has chondroprotective properties attributable to suppression of IL-1â mediated promatrix metalloproteinase production and proteoglycan release.

In patients with osteoarthritis of the knee, aceclofenac decreases pain, reduces disease severity and improves the functional capacity of the knee. It reduce joint reduces inflammation, pain intensity and the duration of morning stiffness in patients with rheumatoid arthritis. The duration of morning stiffness and pain intensity are reduced and spinal mobility improved, by aceclofenac in patients with ankylosing spondylitis.

Pharmacokinetics

Aceclofeanc is rapidly and completely absorbed after oral administration. Peak plasma concentrations are reached to 1 to 3 hours after an oral dose. The drug is highly protient bound (99%). The presence of food does not alter the extent of absorption of aceclofenac but the absorption rate is reduced. The plasma concentration of aceclofenac was approximately twice that in synovial fluid after multiple doses of the drug in patients with knee pain and slynovial fludie effusiton.

Aceclofeance is metabolized to a major metabolite, 4'-hydroxyaceclofenac and to a number of other metabolites including 5-hydroxyaceclofenac, 4'- hydroxydiclofenac, diclofenac and 5-hydroxydiclofenac. These other metabolites account for the fate of approximately 20 % of each does of aceclofenac. Renal excretion is the main route of elimination of aceclofenac with 70 to 80% of an administered does found in the urine, mainly as the glucuronides of aceclofenac and its metabolites. Of each dose of aceclofenac, 20% is excreted in the faeces. The plasma elimination half-life of the drug is approximately 4 hours.

(2) Initial HPLC Condition

Mobile phase: Buffer-ACN (68: 32, v/v)

Buffer: 0.01M ammonium acetate buffer with 2 ml triethylamine, pH 6.5 with glacial Acetic acid.

Column: Phenomenex, C18 (250 mm x 4.6 mm i.d., 5μ particle size)

Flow rate: 1 ml/min

Wavelength: 270 nm

Injection volume: 20 μl

Diluent: Water: ACN (50: 50, v/v)

Sample Preparation

(I) Standard Preparation

Standard solution containing tramadol hydrochloride (0.0375 mg/ml) and aceclofenac (0.100 mg/ml) were prepared by dissolving 18.75 mg tramadol hydrochloride and 50 mg aceclofenac in 50 ml volumetric flask by diluent (stock standard solution). Pipette out 5 ml stock solution into 50 ml volumetric flask and dilute up to mark with diluent (standard solution).

(II)   Test Preparation

Twenty tablets were weighed and the average tablet weight was determined. Tablets were crushed by mortar and pastel. Tablet powder was weighed equivalent to five times of average weight and transfer in to 500 ml volumetric flask. About 50 ml methanol and 300 ml mobile phase was added and sonicated for of 20 min. time interval with intermittent shaking. Content was brought back to room temperature and dilute to volume with diluent (stock test solution). The stock solution was filtered through 0.45 μm nylon syringe filter. Pipette out 5ml filtered stock solution in to 50 ml volumetric flask and dilute with diluent (test solution). The concentration obtain was 0.0375 mg/ml of tramadol hydrochloride and 0.100 mg/ml of aceclofenac.

(3) Literature Survey

(1) **P. Musmade, G. Subramanian, K. K.Srinivasan** have developed a simple HPLC method for quantification of aceclofenac in rat plasma. Ibuprofen was used as an internal standard. Separation was carried out on reversed-phase C18column (250 mm x 4.6 mm, 5 μm particle size) and the column effluent was monitored by UV detector at 282 nm. The mobile phase used was methanol-0.3% (v/v) triethylamine, pH 7.0, (60:40, v/v) at a flow rate of 1.0 ml/min. The method was applied for pharmacokinetic study of aceclofenac in rats.

(2) **A. Zinellu, C. Carru, S. Sotgia, E. Porqueddu, P. Enrico, L. Deiana** have reported a fast-free zone capillary electrophoresis method for the simultaneous determination of aceclofenac and diclofenac in human plasma.

A separation was achieved using a 40 cm × 75 μm uncoated silica capillary, 300 mM/L sodium borate buffer, 200 mM/l N-methyl-d-glucamine, pH 8.9, in about 3 min. method is efficient mean for the comprehensive determination of aceclofenac and diclofenac in human plasma when pharmacokinetics studies are required.

(3) **Y. Jin, H. Chen, S. Gu, F. Zeng** has established a reversed phase HPLC method for the determination of aceclofenac in human plasma.

Chromatography was performed on a C18 column with methanol-0.1 M/L ammonium acetate, pH 6.0, (7:3, v/v) as the mobile phase. The flow rate was 1.0 ml/min. The UV- detector was set at 275 nm. This method can be used for clinical pharmacokinetic study of aceclofenac.

(4) **N. Y. Hasan, M. Abdel-Elkawy, B. E. Elzeany, N. E. Wagieh** have established five new methods for the determination of aceclofenac in the presence of its degradation product diclofenac. Method-A utilizes third derivative spectrophotometry at 242 nm. Method-B was RSD$_1$ spectrophotometric method based on the simultaneous use of the first derivative of ratio spectra and measurement at 245 nm. Method-C was a Ph induced difference (ÄA) spectrophotometry using UV measurement at 273 nm. Method-D was a spectrodensitometric one, which depends on the quantitative densitometric evaluation of thin layer chromatogram of aceclofenac at 275 nm. Method-E was RP-HPLC that depends on using methanol- water (60:40, v/v) as mobile phase at a flow rate of 1 ml/min and UV detection at 275 nm. The methods could be applied for the analysis of the drug in its pharmaceutical formulation.

(5) **B. Hinz, D. Auge, T. Rau, S. Rietbrock, K. Brune, U. Werner** have been reported a method for the simultaneous determination of aceclofenac and three of its metabolites in human plasma by HPLC. The analytes

were separated using an acetonitrile-phosphate buffer gradient at a flow rate of 1 ml/min and UV- detection at 282 nm. The developed procedure was applied to assess the pharmacokinetics of aceclofenac and its metabolites.

(6) **N. H. Zawilla, M. A. A. Mohammad, N. M. El Kousy, S. M. El-Moghazy Aly** have reported three sensitive and reproducible methods for quantitative determination of aceclofenac in pure form and in pharmaceutical formulation. The first method is based on Beer's law. Absorption measurements were carried out at 665.5 nm. The other two methods are high performance liquid chromatography and densitometric methods by which the drug was determined in the presence of its degradation products.

(7) **H. S. Lee, C. K. Jeong, S. J. Choi, S. B. Kim, M. H. Lee, G. Il Ko and D.H. Sohn** have developed a narrow bore HPLC with column-switching for the simultaneous determination of aceclofenac and diclofenac from human plasma samples. Plasma sample (100 µl) was directly introduced onto a Capcell Pak MF Ph-1 column (20 mm × 4 mm i.d.) primary separation was occurred to remove proteins and concentrate target substances using acetonitrile- 0.1 M potassium phosphate, pH 7, (14:86, v/v). The drug molecules eluted from MF Ph-1 column were focused in an intermediate column (35 mm × 2 mm i.d.) by the valve-switching step. The substances enriched in intermediate column were eluted and separated on the narrow bore phenyl–hexyl column (100 mm × 2 mm i.d.) using acetonitrile-0.02M potassium phosphate, pH 7, (33:67, v/v).

(8) **X. Q. Liu, X. J. Chen, L. H. Zhao, J. H. Peng** have reported HPLC assay method for the determination of aceclofenac in plasma and its pharmacokinetics in dogs.

6. Experimental

6.1. Development and Optimization of the HPLC Method

In the presence work, an analytical method based on LC using UV detection was developed and validated for assay determination of aspirin and clopidogrel in tablet formulation. The analytical conditions were selected, keeping in mind the different chemical nature of aspirin and clopidogrel. The development trials were taken by using the degraded sample of each component was done, by keeping them in various extreme conditions.

The column selection has been done on the basis of backpressure, resolution, peak shape, theoretical plates and day-to-day reproducibility of the retention time and resolution between aspirin and clopidogrel peak. After evaluating all these factors, C8 (2) (250 mm 4.6 mm i.d., 5 µm particle size) column was found to be giving satisfactory results. The selection of buffer based on chemical structure of both the drugs. The acidic pH range was found suitable for solubility, resolution, stability, theoretical plates and peak shape of both components. Best results were obtained with 0.3% orthophosphoric acid solution improved the peak shape of aspirin and clopidogrel. Finally, by fixing 0.3% orthophosphoric acid (v/v) and mobile phase composition consisting of a mixture of 0.3% orthophosphoric acid (v/v)-acetonitrile (65:35, v/v). Optimized mobile phase proportion was provided good resolution between aspirin and clopidogrel and also for degradation product which is generated during force degradation study. For the selection of organic constituent of mobile phase, acetonitrile was chosen to reduce the longer retention time and to attain good peak shape. Figure 3 and Figure 4 represent the chromatograms of standard and test preparation respectively.

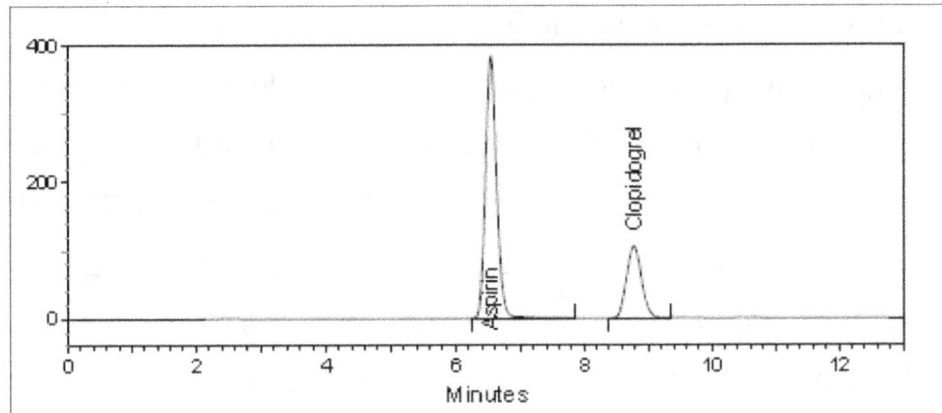

Chromatogram of Standard preparation

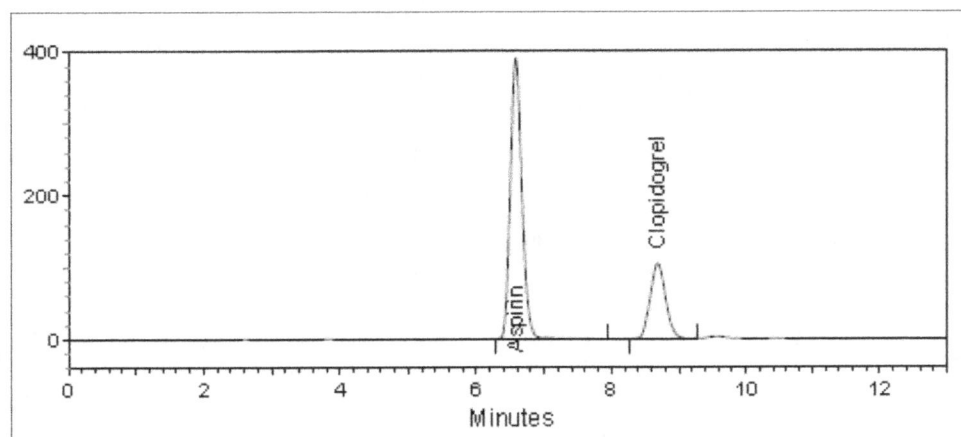

Chromatogram of Test preparation

6.2. Method Validation

6.2.1. *Specificity:* The specificity of the method was determined by checking the interference of placebo with analyte and the proposed method were eluted by checking the peak purity of aspirin and clopidogrel during the force degradation study. The peak purity of the aspirin and clopidogrel were found satisfactory under different stress condition. There was no interference of any peak of degradation product with drug peak.

6.2.2. *Linearity:* For linearity seven points calibration curve were obtained in a concentration range from 0.030-0.120 mg/ml for aspirin and 0.015-0.060 mg/ml for clopidogrel. The response of the drug was found to be linear in the investigation concentration range and the linear regression equation for aspirin was y= 60026378.57x + 51410.11 with correlation coefficient 0.9999 shown in figure, and for clopidogrel was y = 44544414.03x1890.29 with correlation coefficient 0.9999 shown in below mention figure. Where x is the concentration in mg/ml and y is the peak area in absorbance unit. Chromatogram obtain during linearity study were shown in following figures.

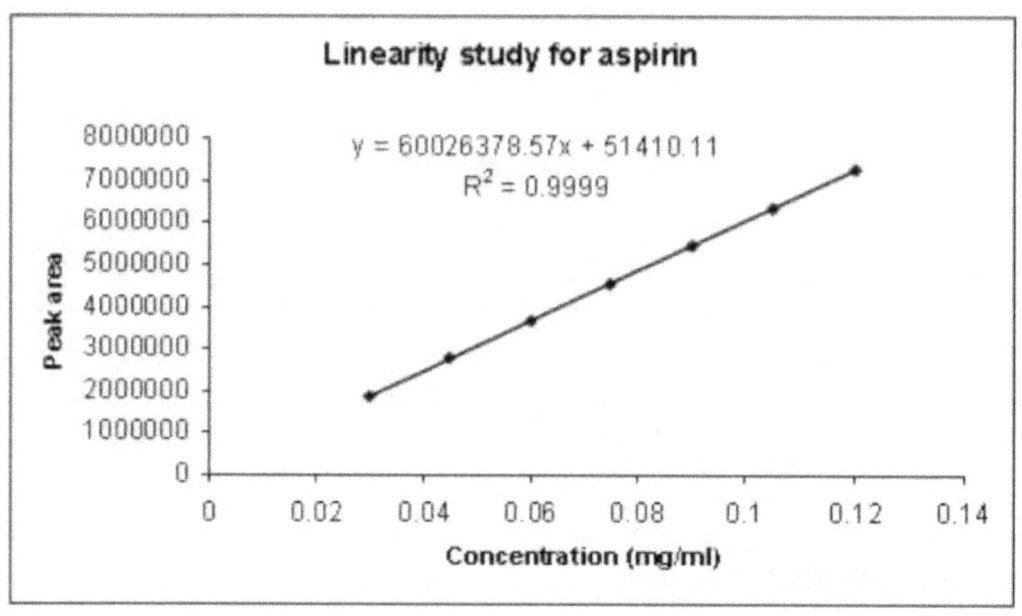

Linearity curve of aspirin

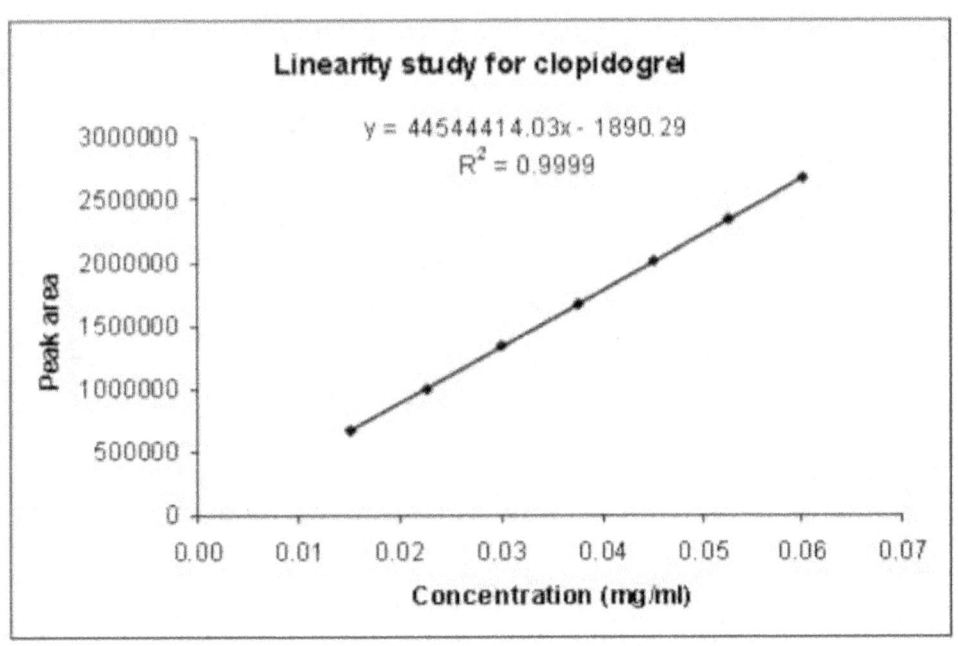

Linearity curve of clopidogrel

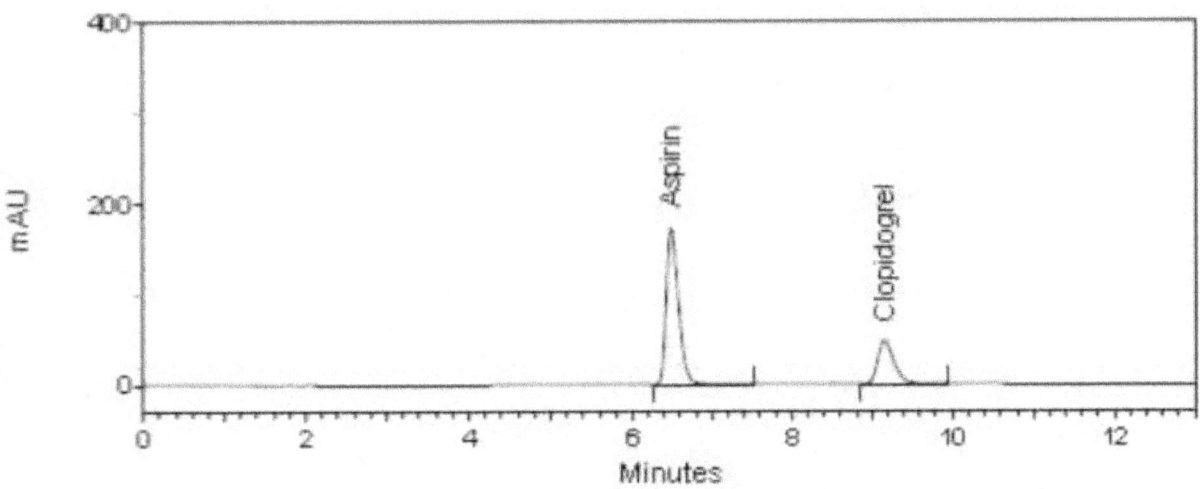

Linearity study chromatogram of level-1

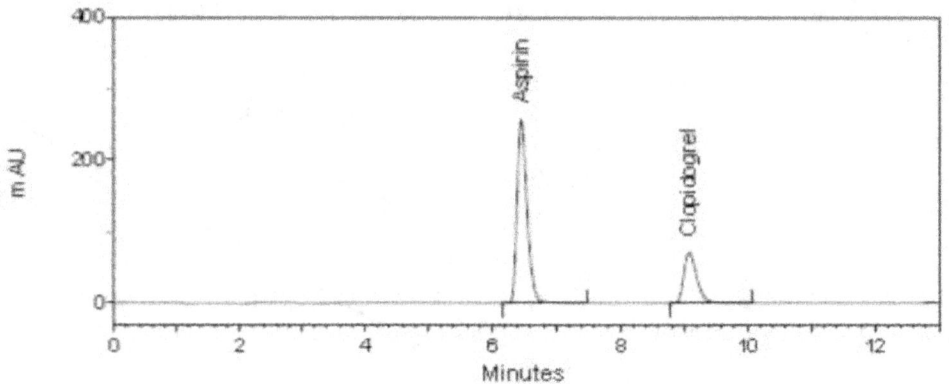

Linearity study chromatogram of level-1

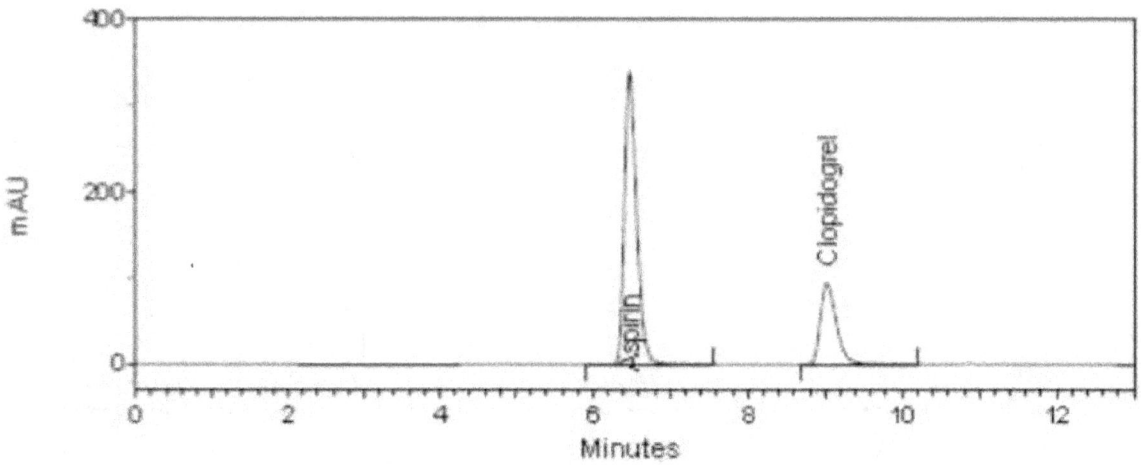

Linearity study chromatogram of level-1

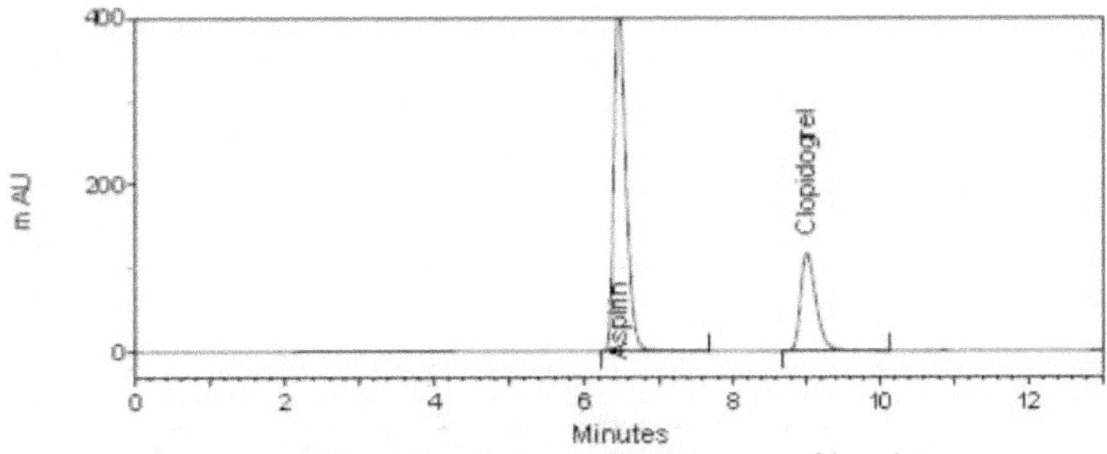

Linearity study chromatogram of level-1

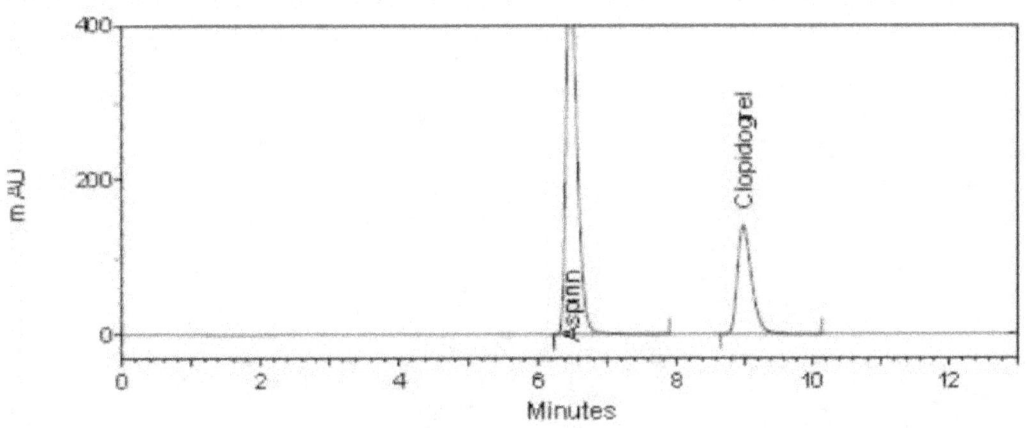

Linearity study chromatogram of level-1

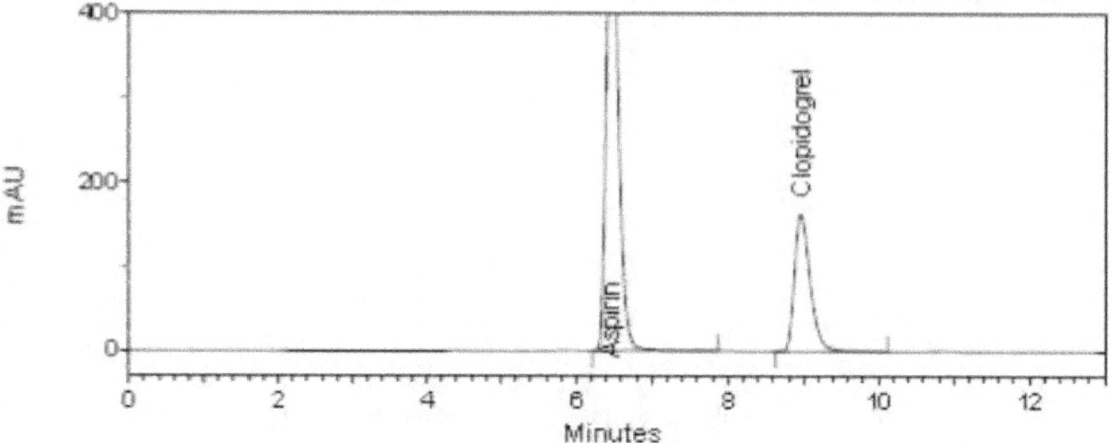

Linearity study chromatogram of level-1

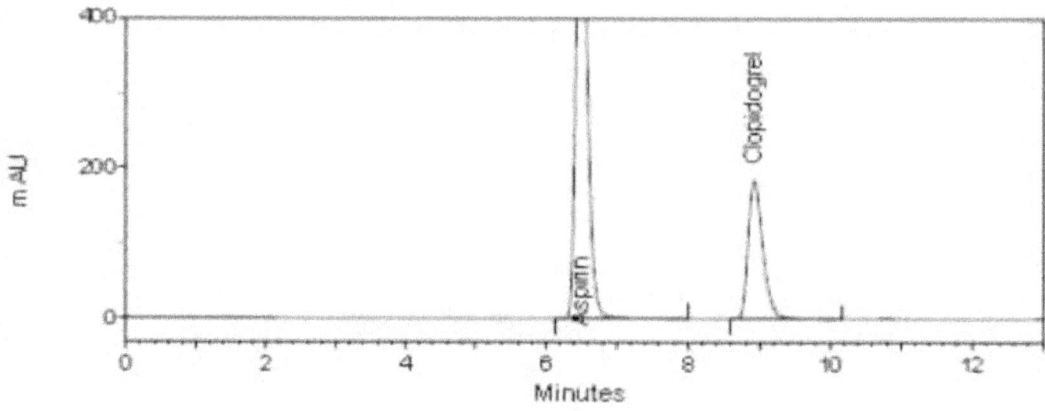

Linearity study chromatogram of level-1

6.2.3. *LOD and LOQ:* The limit of detection and limit of quantification were evaluated by serial dilutions of aspirin and clopidogrel stock solution in order to obtain signal to noise ratio of 3:1 for LOD and 10:1 fro LOQ. The LOD value for aspirin and clopidogrel were found to be 0.05 ppm and 0.15 ppm, respectively and the LOQ value 0.2 ppm and 0.3 ppm, respectively. Chromatogram of LOD and LOQ study were shown in following figures.

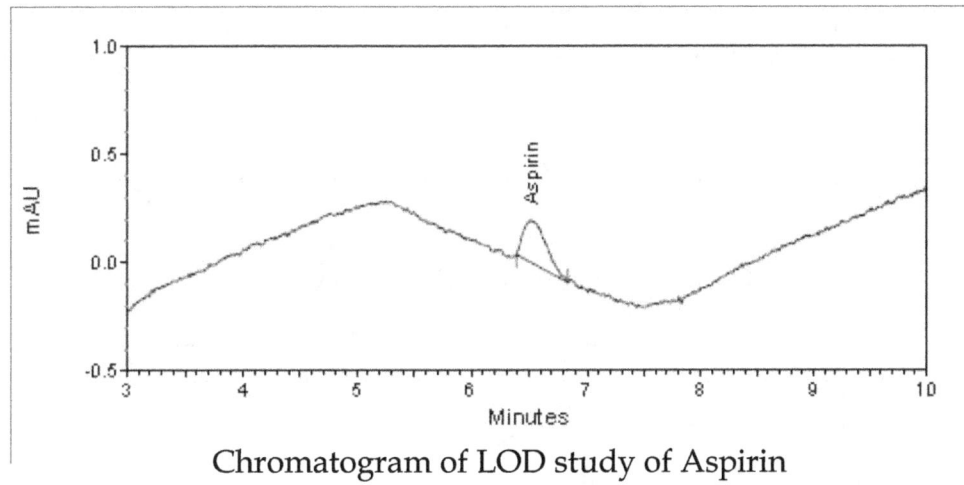

Chromatogram of LOD study of Aspirin

Chromatogram of LOD study of Clopidogrel

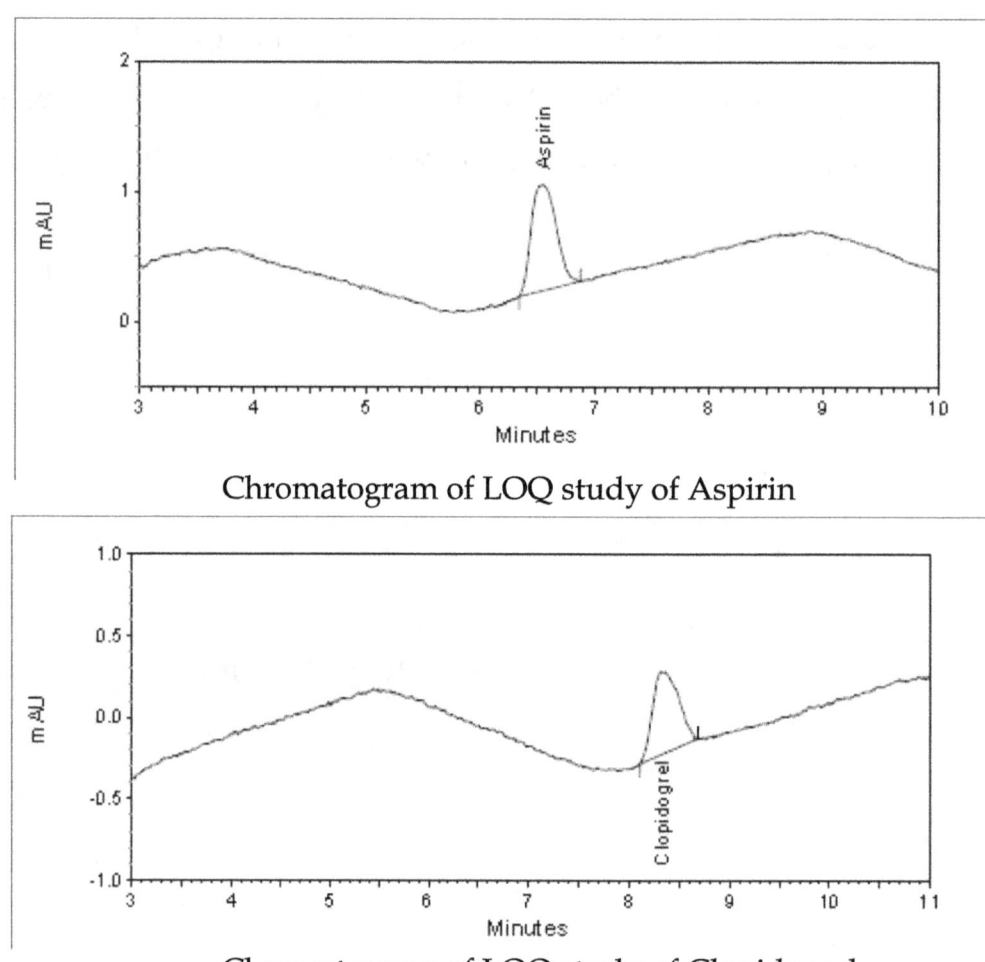

Chromatogram of LOQ study of Aspirin

Chromatogram of LOQ study of Clopidogrel

6.2.4. *Precision:* Precision was investigated using the sample preparation procedure for six real samples of commercial API of clopidogrel bisulphate and Aspirin both.

Method Precision (Intra-day): The precision of the method was evaluated by carrying out five independent assays of Aspirin and Clopidogrel both (100 µg/ml) test sample which is the standard one.

Data obtain from precision experiments are given in below Table for intra-day and inter-day precision study for Aspirin and clopidogrel both. The percentage RSD values for intra-day precision study and inter-day precision study was < 2.0 % for both. This is confirming a good precision.

Result of precision study

Set	Aspirin (%Assay)		Clopidogrel (%Assay)	
	Intraday (n = 6)	Interday (n = 6)	Intraday (n = 6)	Intraday (n = 6)
1	99.1	100.2	99.3	99.6
2	100.0	99.9	98.7	99.6
3	99.6	100.5	98.6	100.1
4	99.5	100.3	99.0	100.1
5	100.3	101.0	100.0	100.6
6	99.1	100.8	99.5	100.7
Mean	99.6	100.5	99.2	100.1
Standard deviation	0.48	0.40	0.53	0.47
% RSD	0.48	0.40	0.53	0.47

6.2.5. *Accuracy:* Recovery of aspirin and clopidogrel were determined at three different concentration levels. The mean recovery for aspirin was 99.12-99.83 % and 98.20-100.35 % for clopidogrel (Table 2). The result indicating that the method was accurate.

Chromatogram obtain during accuracy study were shown in following figures.

Result of accuracy study

	Level (%)	Amount Added Concentration[a] (mg/ml)	Amount Found Concentration[a] (mg/ml)	% Recovery[a]	% RSD[a]
Aspirin	50	0.03751	0.03721	99.22	0.07
	100	0.07497	0.07432	98.12	0.23
	150	0.11250	0.11232	99.83	0.05
Clopidogrel	50	0.01874	0.01840	98.20	0.19
	100	0.03748	0.03695	98.59	0.14
	150	0.05627	0.05647	100.35	0.24

[a] Each value corresponds to the mean of three determinations.

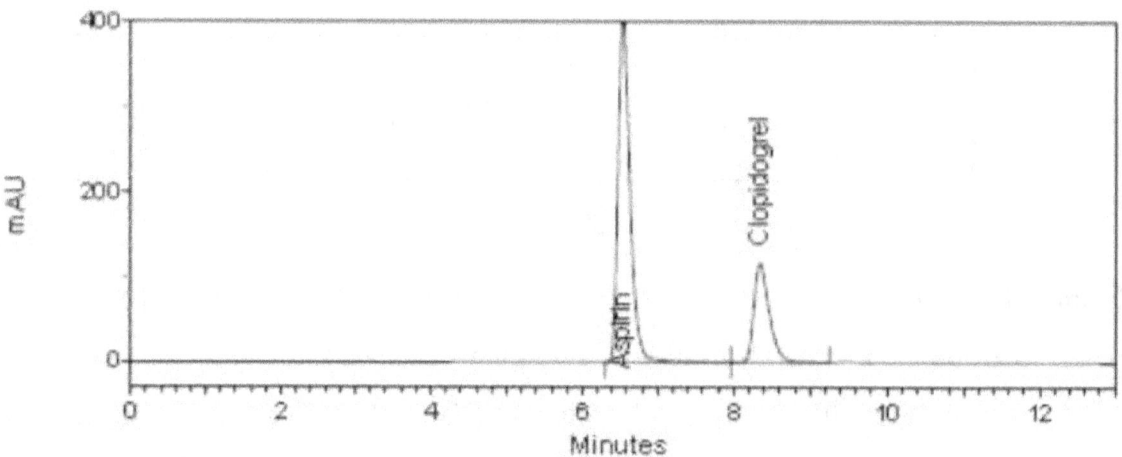

Accuracy study chromatogram of level

56

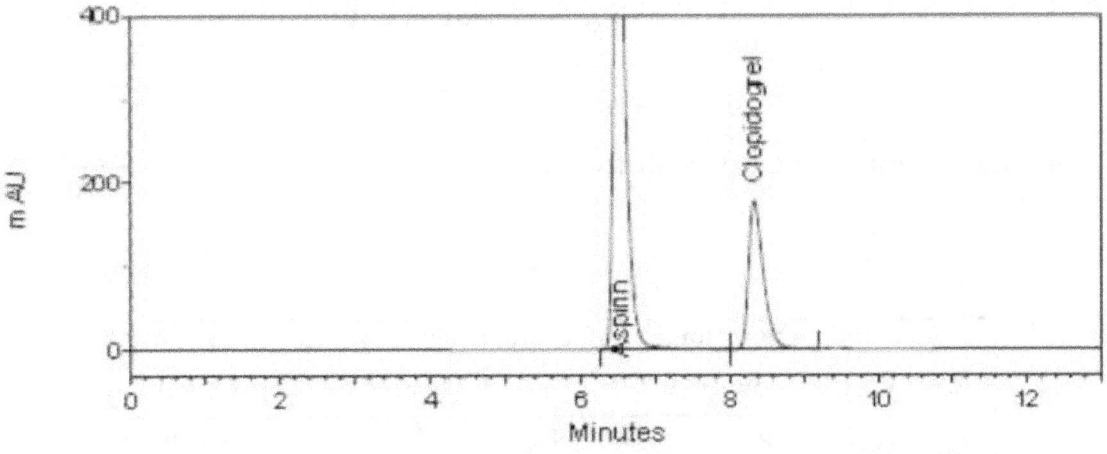

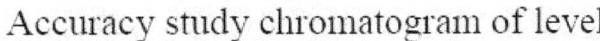

Accuracy study chromatogram of level

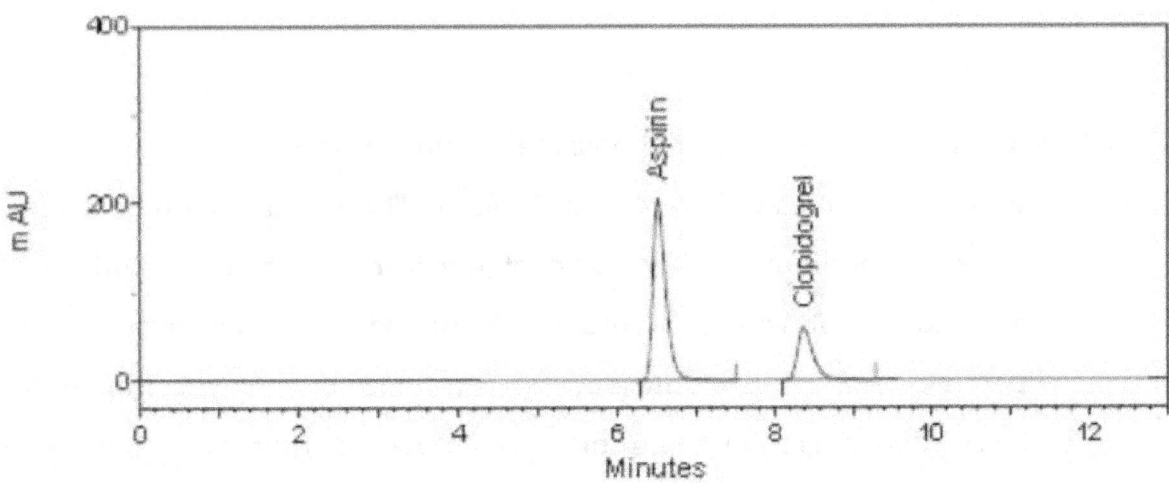

Accuracy study chromatogram of level

6.2.6. *Solution stability study:* Following table shows the results obtain in the solution stability study at different time intervals for test preparation. It was concluded that the test preparation solution was found stable up to 48 h at 2-5 oC and 36 h at ambient temperature with the consideration of < 2.0% in %assay value difference of interval value against initial value.

Evaluation data of solution stability study

Intervals	% Assay for Test Solution Stored at 2 –5 ^{0}C		% Assay for Test Solution Stored at Ambient Temperature	
	Aspirin	Clopidogrel	Aspirin	Clopidogrel
Initial	101.3	100.0	101.3	100.0
12 h	100.9	99.3	100.1	99.5
24 h	100.3	99.6	99.8	99.0
36 h	100.1	99.0	99.8	98.6
48 h	99.9	98.7	98.1	98.7

6.2.7. *Robustness:* The result of robustness study of the developed assay method was established in Table 4 and Table 5. The result shown that during all variance conditions, assay value of the test preparation solution was not affected and it was in accordance with that of actual. System suitability parameters were also found satisfactory; hence the analytical method would be concluded as robust. Chromatogram obtain during robustness study were shown in following figures.

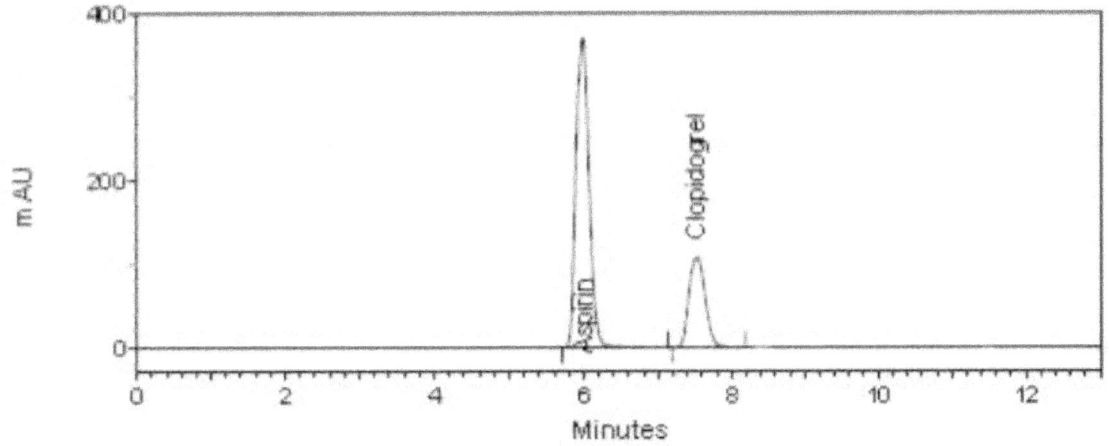

Standard chromatogram [0.3% H₃PO₄-ACN (63:37)

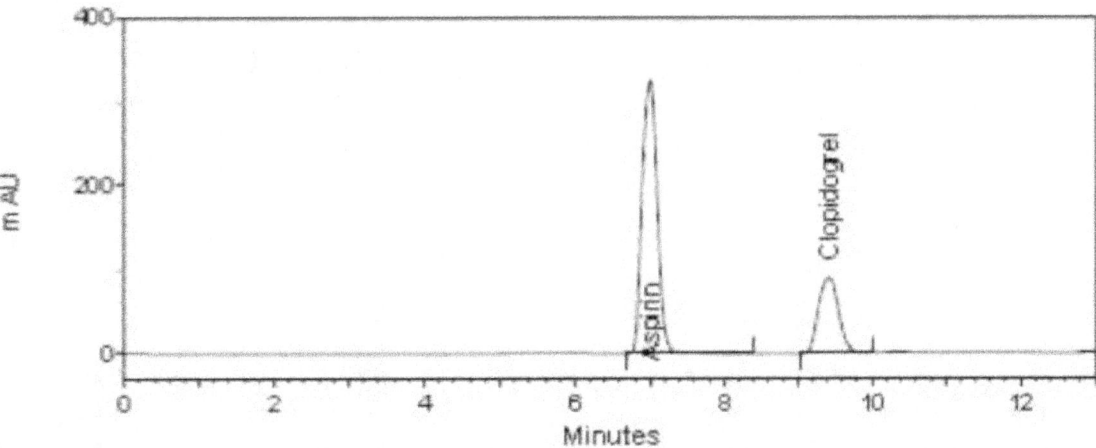

Standard chromatogram [0.3% H₃PO₄-ACN (67:33)

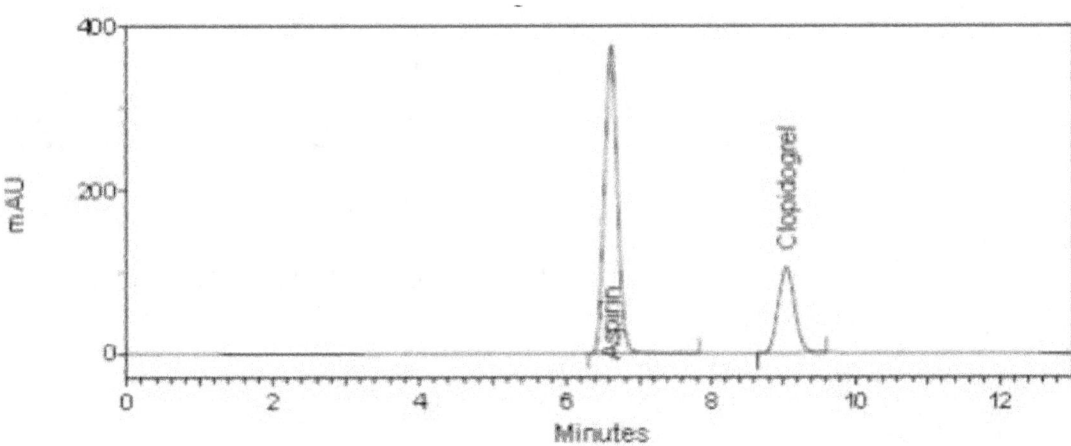

Standard chromatogram [0.28% H₃PO₄-ACN (65:35)

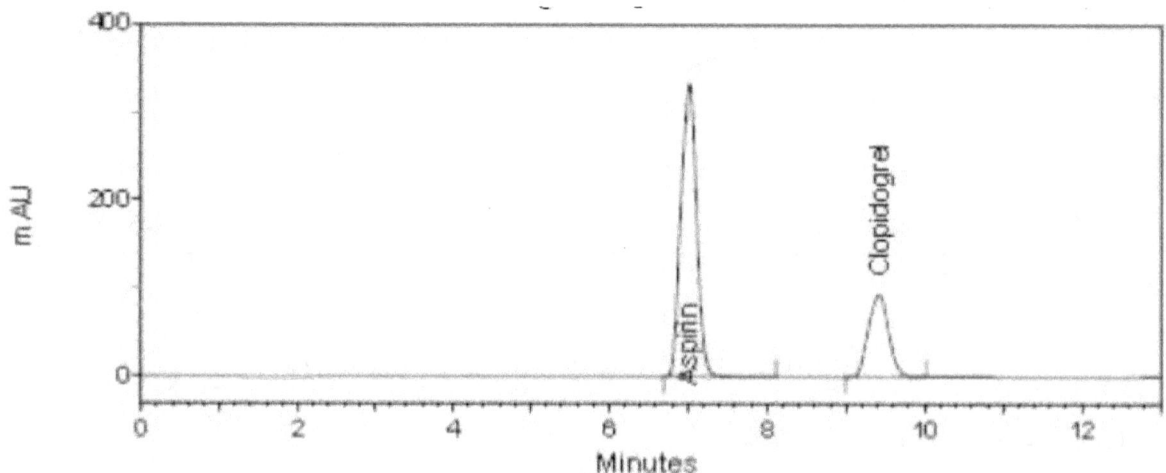

Standard chromatogram [0. 32% H₃PO₄-ACN (65:35)

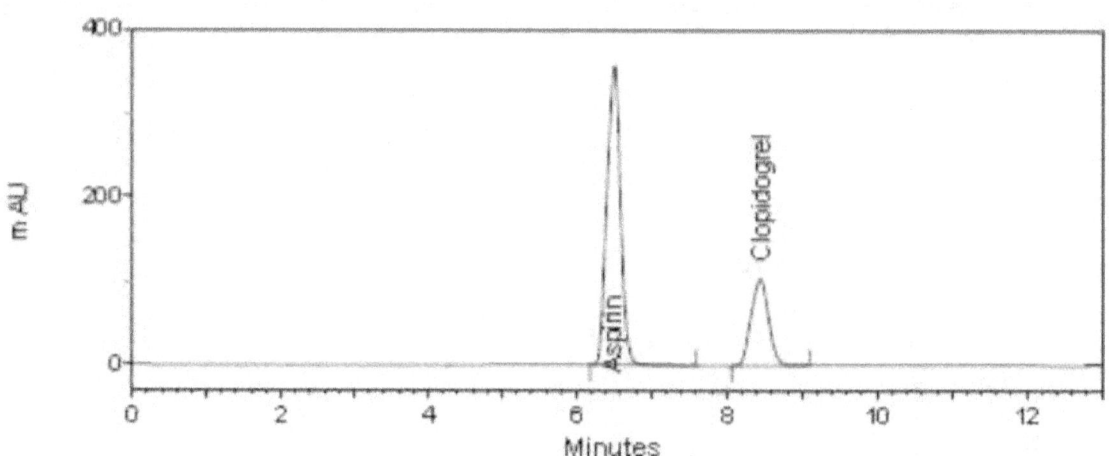

Standard chromatogram (Column change)

60

Evaluation data of robustness study of aspirin

Robust Conditions	% Assay	System Suitability Parameters		
		Theoretical Plates	Asymmetry	% RSD
Flow 0.9 ml/min	100.5	6460	1.05	0.22
Flow 1.1 ml/min	100.3	5661	1.05	0.09
0.28 % H_3PO_4-ACN (65:35, v/v)	100.0	6117	1.00	0.40
0.32 % H_3PO_4-ACN (65:35, v/v)	99.7	5588	1.02	0.30
0.3% H_3PO_4-ACN (63:37, v/v)	100.2	5475	1.12	0.19
0.3% H_3PO_4-ACN (67:33, v/v)	100.1	5838	1.04	0.20
Column change	100.4	5425	1.05	0.34

Evaluation data of robustness study of clopidogrel

Robust Conditions	% Assay	System Suitability Parameters			
		Theoretical Plates	Asymmetry	% RSD	Resolution
Flow 0.9 ml/min	98.5	6975	1.07	0.67	5.75
Flow 1.1 ml/min	100.0	5992	1.06	0.39	5.14
0.28 % H_3PO_4-ACN (65:35, v/v)	98.8	6899	1.03	1.03	6.35
0.32 % H_3PO_4-ACN (65:35, v/v)	99.2	6113	1.03	0.69	5.62
0.3% H_3PO_4-ACN (63:37, v/v)	99.5	5850	1.11	0.70	4.31
0.3% H_3PO_4-ACN (67:33, v/v)	99.0	6185	1.04	0.30	5.71
Column change	100.0	5996	1.04	0.25	4.97

7. Bibliography

1. http:// www. umich.edu/~orgolab/Chroma/chromahis.html

2. From Wikipedia, the free encyclopedia

3. http:// kerouac.pharm.uky.edu/asrg/hplc/history.html

4. http:// www. laballiance.com/la info%5Csupport%5Chplc3.htm

5. Vander Wal S, Snyder LR. *J. Chromatogr.* 225 (1983) 463.

6. *A Practical Guide to HPLC Detection*, Academic Press, San Diego, CA, (1983).

7. Poole CF, Schutte SA. *Contemporary Practice of Chromatography*, Elsevier Amsterdam, (1984) 375.

8. Krull IS. In *Chromatography and Separation Chemistry: Advances and Developments*, Ahuja S. ed., ACS Symposium Series 297, ACS, Washington, DC, (1986) 137.

9. Li G, Szulc ME, Fischer DH, Krull IS. In *Electrochemical Detection in Liquid Chromatography and Capillary Electrophoresis*, Kissinger PT. edn., *Chromatography Science Series*, Marcel Dekker, New York, (1997).

10. Kissinger PT, Heineman WR. eds., *Laboratory Techniques in Electroanalytical Chemistry*, Chaptor 20, Marcel Dekker, New York, (1984).

11. Swarbrick JC, Boylan James, *Encyclopedia of pharmaceutical technology*, Vol. I (1998) 217-224.

12. Lindsay Sandy, *HPLC by open learning*, (1991) 30-45.

13. Lough WJ, Wainer IWW. *HPLC fundamental principles and practices*, (1991) 52-67.

14. Krstulovic AM, Brown PR. *Reversed-Phase High Performance Liquid Chromatography: Theory, Practice and Biomedical Applications*, Wiley, Newyork, (1982).

15. U.S. FDA, Title 21 of the U.S. Code of Federal Regulations: 21 CFR 211- Current good manufacturing practice for finished pharmaceuticals.

16. U.S. FDA - Guidance for Industry (draft) *Analytical Procedures and Methods Validation:* Chemistry, Manufacturing, and Controls and Documentation, (2000).

17. ISO/IEC 17025, General requirements for the competence of testing and calibration laboratories, (2005).

18. International Conference on Harmonization (ICH) of Technical Requirements for the Registration of Pharmaceuticals for Human Use, *Validation of analytical procedures: definitions and terminology*, Q2A, Geneva (1996).

19. International Conference on Harmonization (ICH) of Technical Requirements for the Registration of Pharmaceuticals for Human Use, *Validation of analytical procedures: Methodology*, Q2B, Geneva (1996).

20. U.S. EPA, Guidance for methods development and methods validation for the Resource Conservation and Recovery Act (RCRA) Program, Washington, D.C. (1995).

21. http://www.epa.gov/sw-846/pdfs/methdev.pdf

22. General Chapter 1225, Validation of compendial methods, *United States Pharmacopeia 30*, National Formulary 25, Rockville, Md., USA, The United States Pharmacopeial Convention, Inc., (2007).

23. Hokanson GC. A life cycle approach to the validation of analytical methods during pharmaceutical product development, Part I: The initial validation process, *Pharm Tech*, Sept. (1994) 118–130.

24. Hokanson G.C., A life cycle approach to the validation of analytical methods during pharmaceutical product development, Part II: Changes and the need for additional validation, *Pharm Tech*, Oct. (1994) 92–100.

25. Green JM. A practical guide to analytical method validation, *Anal Chem News & Features*, 1 May (1996) 305A–309A.

26. Wegscheider, Validation of analytical methods, in: *Accreditation and quality assurance in analytical chemistry*, edited by Guenzler H, Springer Verlag and Berlin (1996).

27. Seno S, Ohtake S, Kohno H. Analytical validation in practice at a quality control laboratory in the Japanese pharmaceutical industry, *Accred Qual Assur*. 2 (1997) 140-145.

28. AOAC Peer-Verified Methods Program, Manual on policies and procedures, Arlington, Va., USA (1998). http://www.aoac.org/vmeth/PVM.pdf

29. Winslow PA, Meyer RF. Defining a master plan for the validation of analytical methods, *J Validation Technology*, (1997) 361–367.

30. Breaux J, Jones K, Boulas P. Pharmaceutical Technology, *Analytical Technology and Testing* (2003) 6-13.

31. Huber L, George S. *Diode-array detection in high-performance liquid chromatography*, New York, Marcel Dekker, ISBN 0-8247-4 (1993).